MONOGRAPHS ON THE
PHYSICS AND CHEMISTRY
OF MATERIALS

General Editors

Quantum Theory of Collective Phenomena

G. L. SEWELL
Department of Physics, Queen Mary College, London

CLARENDON PRESS • OXFORD
1986

Oxford University Press, Walton Street, Oxford OX2 6DP

Oxford New York Toronto
Delhi Bombay Calcutta Madras Karachi
Kuala Lumpur Singapore Hong Kong Tokyo
Nairobi Dar es Salaam Cape Town
Melbourne Auckland
and associated companies in
Beirut Berlin Ibadan Nicosia

Oxford is a trade mark of Oxford University Press

Published in the United States
by Oxford University Press, New York

British Library Cataloguing in Publication Data
Sewell, G. L.
Quantum theory of collective phenomena.——
(*Monographs on the physics and chemistry of materials*)
1. *Statistical mechanics* 2. *Quantum field theory*
I. Title II. Series
530.1′33 *QC*174.45

ISBN 0-19-851371-2

Library of Congress Cataloging in Publication Data
Sewell, G. L.
Quantum theory of collective phenomena.
(*Monographs on the physics and chemistry of materials*)
Bibliography: *p.*
Includes index.
1. *Quantum theory.* 2. *Statistical thermodynamics.*
3. *Phase transformations* (*Statistical physics*)
I. Title. II. Title: *Collective phenomena*
*QC*174.12.*S*48 1985 530.1′2 85-15522
ISBN 0-19-851371-2

Typeset and Printed in Northern Ireland by
The Universities Press (*Belfast*) *Ltd*

Preface

The macroscopic properties of matter are governed by quantum mechanical processes that are collective, in that they involve the co-operation of enormous numbers of particles. Correspondingly, the quantum theory of macroscopic phenomena requires concepts, such as those of order and entropy, that represent collective effects in 'very large' assemblies of particles. It is therefore radically different from the quantum theory of atoms and small molecules, where such concepts have no relevance.

Important developments in the quantum theory of macroscopic, or collective, phenomena have ensued from the discovery that the idealization of many-particle systems as infinite can reveal some of their intrinsic properties that would otherwise be masked by finite-size effects. The essential reason for this may be traced to the fact that this idealization permits qualitative distinctions between the descriptions of matter at the macroscopic and microscopic, or global and local, levels, whereas the corresponding distinctions for finite systems are merely quantitative. Thus, for example, an infinite system, unlike a finite one, admits inequivalent representations of its observables, corresponding to macroscopically different classes of states, such as those belonging to different thermodynamic phases. Consequently, it emerges that the model of an infinite system provides the conceptual structure needed for a theory of phase transitions, characterized by spontaneous symmetry changes as well as thermodynamical singularities. It also provides the framework for theories of irreversible processes, free from Poincaré cycles, and metastable states, characterized by stability of a lower grade than that of thermal equilibrium.

The object of this book is to provide a systematic approach to the quantum theory of collective phenomena, based principally on the model of infinite systems. The book is addressed to physicists and chemists who are interested in understanding the scope of this approach, and also to mathematicians who may wish to study the structure and physical relevance of the model. Throughout the book, I have aimed to keep the mathematics as simple as possible, without sacrificing rigour. The book is thus designed to be readable on the basis of a knowledge of standard quantum theory and statistical mechanics, and of the essentials of mathematical analysis and vector space theory. Any additional mathematics required here, mainly elementary functional analysis, will be expounded in self-contained form, either in the main text or in the Appendices: for example, an Appendix to Chapter 2 is devoted to an exposition of the elements of Hilbert space theory.

The book consists of three parts. Part I is an exposition of the generalized form of quantum theory of both finite and infinite systems. Part II consists of a general formulation of statistical thermodynamics within the framework of Part I. This contains what I believe to be a new derivation of thermodynamics with phase structure (in Chapter 4), which has been obtained by incorporating conserved global observables into the model of infinite systems. Part III provides a treatment of the phenomena of phase transitions, metastability and the generation of ordered structures far from equilibrium, within the framework of Parts I and II. This serves to co-ordinate the theory of these phenomena, placing the results obtained by various methods in a general scheme. It will be seen that, while some of these results can be obtained by traditional methods, there are also some whose very conception requires the idealization of infinite systems.

Since a number of the topics treated in this book are enormous subjects in themselves, I have had to be highly selective in my choice of material. The choice made here, while inevitably dependent upon my own interests, has been governed by the aim of providing a coherent and relatively simple approach to the theory of collective phenomena.

London G. L. S.
January 1985

Acknowledgements

It is a great pleasure to thank Professor H. Fröhlich, Editor of this series of monographs, for inviting me to write this book and for various constructive suggestions and criticisms during its preparation. I am also grateful to Professor O. Penrose for some very useful criticisms of a preliminary draft of the first part of the book. Finally, I would like to thank Mrs. M. Puplett for her patience and efficiency in typing the manuscript.

Contents

Part I

The generalized quantum mechanical framework

1

Introductory discussion on the quantum theory of macroscopic systems

Macroscopic systems enjoy properties that are qualitatively different from those of atoms and molecules, despite the fact that they are composed of the same basic constituents, namely nuclei and electrons. For example, they exhibit phenomena such as phase transitions, dissipative processes, and even biological growth, that do not occur in the atomic world. Evidently, such phenomena must be, in some sense, *collective*,† in that they involve the cooperation of enormous numbers of particles: for otherwise the properties of macroscopic systems would essentially reduce to those of independent atoms and molecules.

The problem of how macroscopic phenomena arise from the properties of the microscopic constituents of matter is basically a quantum mechanical one. That quantum, rather than classical, mechanics is essential here is evident from the great wealth of phenomena in which quantum effects operate on the macroscopic scale. For example, the Third Law of Thermodynamics is a quantum law, the stability of matter‡ itself is a quantum phenomenon; while the physical processes of Josephson tunnelling§ and laser radiation,¶ as well as certain biological ones,|| are characterized in terms of 'macroscopic wave-functions' of a purely quantum nature.

The quantum theory of macroscopic systems is designed to provide a model relating the bulk properties of matter to the microscopic ones of its constituent particles. Since such a model must possess the structure needed to accommodate a description of collective phenomena, characteristic of macroscopic systems only, it is evident that it must contain concepts that are *qualitatively* different from those of atomic physics.

In order to see the nature of the problems involved here, we start by noting that a macroscopic system is composed of an enormous number, e.g. 10^{24}, of interacting particles of one, or possibly several, species. At a microscopic level, therefore, its properties are governed by the Schrödinger equation for this assembly of particles. However, in view of the huge

† The idea of collective behaviour of many-particle systems was first explicitly introduced by Bohm and Pines (1951).

‡ See Dyson and Lenard (1967, 1968); Lieb and Thirring (1975); Lieb (1976).

§ See Josephson (1964).

¶ See Graham and Haken (1970).

|| See H. Fröhlich (1969).

number of particles in the system, this equation is fantastically complicated: indeed its extreme complexity represents an essential part of the physical situation, being closely connected with the 'molecular chaos' that is basic to statistical mechanics. It is this complexity that makes the many-body problem of extracting physically relevant information from the Schrödinger equation radically different from anything encountered in atomic physics. For, as cogently argued by H. Fröhlich (1967, 1973), it imposes a situation where detailed solution of the Schrödinger equation would be too complex to even be contemplated† and where, consequently, the essential role of this equation in the many-body problem is that of a key to interrelationships between appropriate macroscopic variables, as in thermodynamics or hydrodynamics. This signifies that the many-body problem should be cast, *ab initio*, in both *macroscopic and microscopic* terms. One can also see this from the empirical fact that the phenomenological properties of a macroscopic many-particle system are determined not only by its microscopic constitution, but also by its thermodynamic phase or, more generally, by the macroscopic constraints imposed on it: for example, a system in the solid phase lacks the hydrodynamical properties of a liquid.

What is needed, then, is a quantum mechanical model that admits precise mathematical description in both macroscopic and microscopic terms. As a prerequisite for this, we require clear-cut characterizations of macroscopic systems and variables, and this poses a problem, since it is not a priori evident how large an assembly of particles must be before it can be considered to be macroscopic. However, an essential clue to the characterization of macroscopicality is that, at an empirical level, the hallmark of macroscopic objects is that their intensive properties, e.g. their equations of state, are independent of their sizes. This indicates that the intensive properties of an assembly of particles of a given species tend to definite limits when the number of particles tends to infinity at constant density; and that a real macroscopic system of these particles is one that is sufficiently large for its intensive properties to be experimentally indistinguishable from these limiting ones. Furthermore, at the statistical mechanical level, it has been proved‡ that intensive properties of many-particle systems do indeed converge to definite limits, as their sizes tend to infinity, under very general conditions on their interactions.§

These considerations lead naturally to a model, in which a macroscopic system is *idealized* as an infinite assembly of particles, whose density is

† This is not merely a technical point since, even if one could solve the Schrödinger equation with the aid of a supercomputer, its solution would surely be so complicated as to be unintelligible; and the problem of extracting physically relevant information from it would presumably be no simpler than the original one.

‡ See Ruelle's book (1969*a*).

§ These conditions exclude gravitational systems, whose large-scale limiting properties are of a different nature (cf. Hertel and Thirring 1971).

finite. It emerges that this model, which has been extensively studied† in the last two decades, possesses just the structures needed for a theory of collective phenomena, exposing in sharp relief certain intrinsic properties of matter that would otherwise be masked by finite-size effects. The model is defined (cf. Chapter 2) so that its observables and states reduce, in each bounded region, to those of a corresponding finite system of particles there; while its dynamics and intensive thermodynamic potentials are defined as infinite volume limits of those of a finite system of particles of the given species. Thus, the model is constructed as an infinite volume limit of that of a large, but finite, system of the given species of particles.

Let us now briefly indicate how this model provides the natural setting for a systematic theory of collective phenomena. Turning first to the thermodynamic properties of matter, the passage to the infinite volume limit serves to simplify the forms of the thermodynamic potentials by eliminating contributions, due to surface and other finite-size effects, whose extreme smallness, in real macroscopic systems, is concealed by the complexity of their mathematical forms. In particular, it leads to a very important gain for the theory of phase transitions. For, whereas the thermodynamic potentials of a finite system are perfectly smooth,‡ their infinite volume limits can possess singularities, and are known to do so for various models.§ Thus, it is only by passing to the infinite volume limit that one can characterize phase transitions by singularities in the thermodynamic potentials. Here, we emphasize that the passage to this limit does not introduce anything spurious into the theory, but simply sharpens huge gradients into the forms of singularities, which they simulate so closely as to be experimentally indistinguishable from them. From a theoretical standpoint, this constitutes a great conceptual and methodological gain, since singularities may be revealed by certain qualitative features of a model, whereas the detection of huge gradients requires detailed computations that may be too complicated to be either feasible or enlightening.

The model of an infinite system also introduces physically relevant new structures into the theory of both equilibrium and non-equilibrium properties of matter, that go far beyond thermodynamics. A key to these structures is the fact that the model admits a clear-cut distinction between local and global variables, the former referring to the bounded spatial regions of the system and the latter to the whole of it. It therefore permits the natural step of designating as *macroscopic* variables the global intensive quantities given by space-averages of local observables – examples of these are the densities of mass and of energy of the system. Furthermore, these

† See, for example, the books by Ruelle (1969*a*), Emch (1972), Dubin (1974), Bratteli and Robinson (1979, 1981), and Thirring (1980).

‡ See Lebowitz (1968).

§ A useful general reference to models with phase transitions is provided by various articles in the book by Domb and Green (1972).

variables may be used to classify the states of the model with respect to their macroscopic properties in a way that would not be possible for finite systems. For the observables of an infinite system generally have an infinity of inequivalent representations,† each corresponding to a class of macroscopically equivalent but microscopically different states, whereas the observables of a finite system have but one‡ irreducible representation. Hence, a state of an infinite system may be defined macroscopically by a representation and microscopically by a vector or density matrix in the representation space.

Thus, the model possesses sufficient structure to admit description in the macroscopic and microscopic terms needed for a theory of collective phenomena. In the following chapters, we shall show how it may be employed to obtain a general derivation of thermodynamics with phase structure (Chapter 4) and to provide the framework for theories of phase transitions, irreversibility, and metastability. In particular, we shall see that it admits theories of phase transitions, characterized not only by thermodynamical singularities, but also by symmetry breakdown corresponding to a certain 'macroscopic degeneracy' (Chapters 2–5); of critical phenomena (Chapter 5); of metastable states, characterized by a limited stability of a local, rather than global, kind (Chapters 3, 6); of irreversible processes, free from Poincaré cycles (Chapter 7); and of the generation of ordered structures, far from equilibrium, by such processes (Chapter 7).

† A simple demonstration of this is given in Chapter 2 (§2.3). For a general treatment, see Emch's book (1972) and references given there.

‡ See Von Neumann (1931).

2

The generalized quantum mechanical framework

2.1. Introduction

The quantum theory of infinite systems, on which we shall base our treatment of collective phenomena, is a non-trivial generalization† of traditional quantum mechanics.‡ In fact, as we shall presently see, it contains physically relevant structures, that do not occur in the quantum theory of finite systems.

In order to elucidate the essential difference between the quantum theories of finite and infinite systems, we start by describing the states and observables of an arbitrary system in the most basic terms. Thus, an observable is a function of the coordinates and momenta of the particles of the system, which may, in principle, be measured. A state, on the other hand, is a statistical quantity, which serves to determine the expectation values of the observables, should any of them be measured. In other words, if A is an observable and ρ a state, then there is a well-defined quantity, $\rho(A)$ representing the expectation value of A when the state of the system is ρ. Hence we may describe the states, in a convention-free manner, as functionals of the observables, which yield their expectation values under specified conditions.

Now, in the standard quantum theory of finite systems, the observables, A, and the pure states, ρ_ψ, are represented by self-adjoint operators, $\hat{A}$, and normalized vectors, ψ, in a Hilbert space, $\mathscr{H}$, in such a way that $\rho_\psi(A) = (\psi, \hat{A}\psi)$. This representation is based on the condition that the operators $\hat{q}$ and $\hat{p}$, corresponding to a coordinate and its conjugate momentum, respectively, satisfy the canonical commutation relation $[\hat{q}, \hat{p}] = i\hbar I$. The full significance of this relation is evinced by a remarkable theorem of Von Neumann (1931), which establishes that all irreducible representations which satisfy it are equivalent to that of Schrödinger, in which $\hat{p}$ is the differential operator $-i\hbar(\partial/\partial q)$ and $\hat{q}$ is the multiplicative operator, q, acting on the square-integrable functions of the particle

† This generalization stems from formulations of quantum theory by Segal (1947), Haag and Kastler (1964), and Kadison (1965). An authoritative survey of its development may be found in Emch's book (1972).

‡ Standard treatises on traditional quantum mechanics of finite systems are provided by the books of, say, Dirac (1958), Schiff (1968), and, on a rigorous level, Von Neumann (1955) and Jauch (1968).

configuration. In fact, it is this theorem that justifies the standard practice of employing the Schrödinger representation, since it ensures that it leads to no loss of generality.

The situation changes radically, however, when one comes to infinite systems, and the terms of Von Neumann's theorem are then no longer applicable. In fact, as we shall presently see, the observables of an infinite system generally admit an enormous variety of *inequivalent representations*,† corresponding to globally, i.e. macroscopically, different classes of states. This leads to a picture in which a state is specified macroscopically by a representation and microscopically by a vector, or density matrix, within the relevant representation space. The dichotomy thereby obtained between the macroscopic and microscopic descriptions of the system provides just the structure needed for a quantum theory of collective phenomena, and thus for the treatment of basic problems in statistical mechanics. For example, it provides the framework for a theory of *phase transitions*, in which a pure phase corresponds to a representation, determined by the relevant thermodynamical constraints, and transitions between phases are characterized not only by singularities in thermodynamic functions, but also by changes in macroscopic structure, e.g. of symmetry.

We shall devote the present chapter to a general formulation of the quantum theory of both finite and infinite systems, so as to provide the framework for the subsequent treatment of collective phenomena. We start, in §2.2, by summarizing the standard quantum theory of finite systems and then casting it into an algebraic form which may be generalized to infinite systems. In §2.3, we provide a simple pedagogical example, which demonstrates how an infinite system can support inequivalent representations, corresponding to macroscopically different classes of states, and we briefly discuss their relevance to the description of thermodynamic phases. In §2.4, we provide the general formulation of infinite systems. This is based essentially on the idea that the states and observables of an infinitely extended system reduce, in each bounded region, to those of a corresponding finite system. Here, the observables are described in terms of their algebraic structure, governed by the canonical commutation relations, rather than on any particular representation of them; while the states are functionals of the observables which reduce locally to states of a finite system. Moreover, the states may be classified in terms of associated representations of the observables. For, as in the particular model of §2.3, the density matrices in a representation space correspond to a family of macroscopically equivalent, but microscopically different, states. We formulate the dynamics of the infinite system in terms of the time-dependent

† The existence of inequivalent representations of the observables of an infinite system has been known for some time (cf. Haag 1961). The correspondence between a representation and a class of macroscopically equivalent class of states has been particularly well exploited by Hepp (1972) in connection with the quantum theory of measurement.

expectation values of the observables, according to a generalized form of the standard Schrödinger or Heisenberg description. In §2.5, we summarize the main conclusions of this chapter.

There are two Appendices at the end of this chapter. The first consists of an account of the rudiments of the theory of Hilbert spaces, which should suffice for the purposes of this book. We suggest that readers who are not conversant with the theory of these spaces might consult this Appendix before proceeding to §2.2. The second Appendix consists of a proof of a technical point arising in §2.4.

2.2. Finite systems

The quantum theory of finite systems may be summarized as follows (cf. Dirac 1958; Von Neumann 1955). There is a one-to-one correspondence between the *observables*, A, of a finite system and the self-adjoint operators, $\hat{A}$, in a certain Hilbert space, $\mathcal{H}$. The *pure states* of the system correspond to the normalized vectors in $\mathcal{H}$, in such a way that the expectation value of an observable, A, for the state represented by the vector ψ, is $(\psi, \hat{A}\psi)$. The *dynamics* is given, in the Schrödinger picture, by the transformations $\psi \to \psi_t \equiv e^{-i\hat{H}t/\hbar}\psi$ of the states, where $\hat{H}$ is the Hamiltonian operator, representing the energy observable of the system. Thus, the expectation value of the observable A at time t, for an evolution from an initial state ψ, is $(\psi_t, \hat{A}\psi_t)$. The model is therefore determined by the Hilbert space representation of its states and observables, and by the specification of its Hamiltonian.

Our objective in this Section is to review the structure of the model, in the light of Von Neumann's uniqueness theorem concerning the representation of its observables, and then to recast it into a form that can be generalized to infinite systems.

2.2.1. *Uniqueness of the representation*

Suppose first, for simplicity, that we have a system of one degree of freedom, whose coordinate and momentum are q and p, respectively. Then, according to quantum theory, q and p may be represented by self-adjoint operators, $\hat{q}$ and $\hat{p}$, in a Hilbert space $\mathcal{H}$, such that

$$[\hat{q}, \hat{p}] = i\hbar I. \tag{1}$$

This is the canonical commutation relation. A particular realization of it is given by the Schrödinger representation, in which $\hat{p}$ is the differential operator $-i\hbar(d/dq)$ and $\hat{q}$ is the multiplicative operator, q, acting on the hilbert space, $\mathcal{H}$, of square-integrable functions of q, i.e.

$$(\hat{q}f)(q) = qf(q); \qquad (\hat{p}f)(q) = -i\hbar \frac{df}{dq}(q). \tag{2}$$

We remark that this representation is irreducible (Von Neumann 1931), i.e. that $\mathcal{H}$ contains no subspace $\mathcal{K}$ whose vectors remain in $\mathcal{K}$ after being acted on by $\hat{q}$ and $\hat{p}$.

The question now arises as to whether there any other representations of the canonical commutation relation (1) – here it suffices to consider irreducible ones, since any representation can be decomposed into these. Von Neumann's theorem (1931) tells us that there are essentially no others. To be precise, the theorem states that, if $\hat{q}'$ and $\hat{p}'$ are self-adjoint operators in a Hilbert space $\mathcal{H}'$, which provide an irreducible representation of the commutation relation

$$[\hat{q}', \hat{p}'] = \mathrm{i}\hbar I', \tag{3}$$

then there is a unitary transformation V of the Schrödinger representation space, $\mathcal{H}$, onto $\mathcal{H}'$ such that

$$\hat{q}' = V\hat{q}V^{-1} \quad \text{and} \quad \hat{p}' = V\hat{p}V^{-1}. \tag{4}$$

This means that the representations of the canonical commutation rule by $(\hat{q}, \hat{p})$ in $\mathcal{H}$ by $(\hat{q}', \hat{p}')$ in $\mathcal{H}'$ are in one-to-one correspondence. We term these representations *unitarily equivalent*, as the correspondence is implemented by a unitary transformation, V. Thus, all irreducible representations of the canonical commutation relation are unitarily equivalent to that of Schrödinger.

The situation is similar for the representation of a Pauli spin $\mathbf{S} = (S_x, S_y, S_z)$ by self-adjoint operators $\hat{\mathbf{S}} = (\hat{S}_x, \hat{S}_y, \hat{S}_z)$ in a Hilbert space, which satisfy the angular momentum commutation reactions

$$[\hat{S}_x, \hat{S}_y] = \mathrm{i}\hbar\hat{S}_z; \qquad [\hat{S}_y, \hat{S}_z] = \mathrm{i}\hbar\hat{S}_x; \qquad [\hat{S}_z, \hat{S}_x] = \mathrm{i}\hbar\hat{S}_y, \tag{5}$$

together with the spin one-half condition, i.e.

$$\hat{\mathbf{S}} \cdot \hat{\mathbf{S}} = \tfrac{3}{4}\hbar^2 I. \tag{6}$$

As is well-known (cf. Wigner 1959), all irreducible representations of the relations (5) and (6) are equivalent to the two-dimensional one given by

$$\hat{\mathbf{S}} = \tfrac{1}{2}\hbar\hat{\boldsymbol{\sigma}}, \tag{7}$$

where the components of $\hat{\boldsymbol{\sigma}}$ are the Pauli matrices

$$\hat{\sigma}_x = \begin{pmatrix} 0 & 1 \\ 1 & 0 \end{pmatrix}, \qquad \hat{\sigma}_y = \begin{pmatrix} 0 & -\mathrm{i} \\ \mathrm{i} & 0 \end{pmatrix}, \qquad \hat{\sigma}_z = \begin{pmatrix} 1 & 0 \\ 0 & -1 \end{pmatrix}. \tag{8}$$

It is convenient to specify these matrices by the following formula for their action on the basis vectors

$\begin{pmatrix} 1 \\ 0 \end{pmatrix} \equiv \psi(1)$ and $\begin{pmatrix} 0 \\ 1 \end{pmatrix} \equiv \psi(-1)$,

$$\hat{\boldsymbol{\sigma}}\psi(s) = (\psi(-s), \mathrm{i}s\psi(-s), s\psi(s)) \quad \text{for} \quad s = \pm 1. \tag{9}$$

Von Neumann's theorem and its analogue for a Pauli spin tell us that all irreducible representations of the canonical commutation relation (1) are equivalent to Schrödinger's, and that all those of the angular momentum conditions (5) and (6) are equivalent to Pauli's. Furthermore, there is no difficulty in generalizing this result to reach the conclusion that the coordinates, momenta, and spins of an arbitrary finite system admit but one irreducible Hilbert space representation, up to unitary equivalence. We assume that the full set of observables of the system may be identified with the self-adjoint operators in an irreducible representation space, $\mathcal{H}$, since these are generated by the functions of the coordinates, momenta, and spins.

2.2.2. *The Hilbert space description*

In view of this identification of the observables with the self-adjoint operators in $\mathcal{H}$, we shall use the symbol, A, to denote both an observable and the operator which represents it.

The pure states of the system are given by the normalized vectors in $\mathcal{H}$, according to the rule that the expectation value of an observable, A, for the state represented by the vector ψ, is $(\psi, A\psi)$. This state therefore corresponds to the *functional* ρ_ψ of the observables given by

$$\rho_\psi(A) = (\psi, A\psi) \equiv \mathrm{Tr}(P(\psi)A), \tag{10}$$

where $P(\psi)$ is the projection operator for ψ.

For a statistical mixture of pure states, given by an orthonormal set of vectors $\{\psi_n\}$, with respective probabilities $\{w_n\}$, the expectation value of an observable A is

$$\rho(A) = \sum_n w_n \rho_{\psi_n}(A), \tag{11}$$

i.e. by (10)

$$\rho(A) = \mathrm{Tr}(\hat{\rho}A), \tag{12}$$

where

$$\hat{\rho} = \sum_n w_n P(\psi_n). \tag{13}$$

The functional ρ thus corresponds to a statistical mixture of pure states, and is generally termed a *mixed state*. Since the w_n's are probabilities, i.e. $w_n \geqslant 0$ and $\sum w_n = 1$, it follows from (13) that $\hat{\rho}$ is a density matrix, i.e. a positive operator of unit Trace (cf. Appendix A). Conversely, any density matrix $\hat{\rho}$ may be expressed in the form (13), with $\{\psi_n\}$ and $\{w_n\}$ its eigenstates and eigenvalues, respectively. This is true even if $\hat{\rho}$ is a one-dimensional projector, i.e. if one of the w_n's is unity and the rest are zero. Consequently, the relation (13) between states, ρ, and density matrices, $\hat{\rho}$, applies to *all states*, whether pure or mixed. Since this relation holds for all the observables A, it follows that the correspondence between ρ and $\hat{\rho}$ is one-to-one. We may therefore use the same symbol, ρ, without circumflex, to denote both a state functional and its associated density matrix, and rewrite eqn (13) as

$$\rho(A) = \mathrm{Tr}(\rho A). \tag{14}$$

The *dynamics* of a conservative system is governed by the Hamiltonian operator, H, representing its energy observable. In a usual way, we may describe the dynamics according to either of the two equivalent schemes, due to Schrödinger and Heisenberg. In Schrödinger's scheme, the evolution of the system over time t corresponds to the transformation $\rho \to \rho_t$ of its states, where, in the density matrix representation

$$\rho_t = U_t^* \rho U_t, \tag{15}$$

and U_t is the unitary operator given by

$$U_t = \mathrm{e}^{\mathrm{i}Ht/\hbar}. \tag{16}$$

The time-dependent expectation value of an observable, A, for an evolution from an initial state ρ, is therefore

$$\rho_t(A) = \mathrm{Tr}(\rho_t A). \tag{17}$$

In Heisenberg's scheme, on the other hand, the dynamics is represented by the transformations $A \to A_t$ of the observables, where

$$A_t = U_t A U_t^*, \tag{18}$$

and the expectation value of the observable A at time t, for an evolution of the system from an initial state ρ, is $\rho(A_t)$. The equivalence of the Schrödinger and Heisenberg schemes follows from the fact that, by eqns (14), (15), (17), and (18), together with the unitarity of U_t,

$$\rho_t(A) = \rho(A_t). \tag{19}$$

We note also that it follows from (16) and (18) that A_t satisfies the Heisenberg equation of motion, given formally by

$$\frac{\mathrm{d}A_t}{\mathrm{d}t} = \frac{\mathrm{i}}{\hbar}[H, A_t]. \tag{20}$$

Note on Poincaré cycles. It follows from (16) that, *if* H has a purely discrete spectrum, then U_t is quasi-periodic in t and therefore, by (19), so too are the time-dependent expectation values, $\rho_t(A)$, of the observables. In other words, the dynamics of the system exhibits *Poincaré cycles* if its Hamiltonian has a purely discrete spectrum. Furthermore, a sufficient condition† for H to have such a spectrum is that $\mathrm{Tr}\,\mathrm{e}^{-\beta H}$ be finite for positive β. Now the

† This condition suffices to establish the discreteness of the spectrum of H since positive operators of finite Trace have purely discrete spectra (cf. Appendix A).

finiteness of this Trace, which is essential for the thermodynamic properties of the system, has been established for finite assemblies of particles, with realistic interactions, enclosed in bounded volumes (cf. Ruelle 1969*a*). Hence, we may conclude that such finite systems generally exhibit Poincaré cycles.

2.2.3. The algebraic description

Having formulated the model in terms of the standard Hilbert space description, we shall now recast it in an algebraic form that can be generalized to infinite systems. For this purpose we first establish that, by basing the model on the *bounded* observables, i.e. on those that correspond to bounded self-adjoint operators in $\mathcal{H}$, we can sharpen its algebraic structure without sacrificing any of its physical content.

Thus we note that, as bounded operators are defined on all vectors in $\mathcal{H}$, it follows that if A and B are bounded observables, then so too are $(A+B)$, $(AB+BA)$ and $i(AB-BA)$. The same would *not* generally be true if A and B were unbounded, for then their domains of definition would not cover the whole of $\mathcal{H}$ (cf. Appendix A), and so it could happen that the limitations of these domains might preclude the definition of observables of the form $(A+B)$ or $(AB+BA)$ or $i(AB-BA)$. Hence, the bounded observables have an algebraic structure that cannot be extended to all the unbounded ones.

On the other hand, we may always express the unbounded observables in terms of the bounded ones. For, if A is an unbounded observable and N is an arbitrary positive number, there is a unique decomposition of A into self-adjoint operators A_N and A'_N, whose spectra lie inside and outside the closed interval $[-N, N]$, respectively; and further A_N converges strongly to A, as $N \to \infty$, on all vectors in the domain of A (cf. Appendix A, eqns A.19–21)), i.e.

$$A = \lim_{N\to\infty} A_N. \tag{21}$$

Hence, by eqns (14) and (21)

$$\rho(A) = \lim_{N\to\infty} \rho(A_N), \tag{22}$$

provided that $\rho(A)$ is well-defined.† Thus, as eqns (21) and (22) express the properties of the unbounded observables in terms of the bounded ones, it follows that we may base the model on the latter observables, without loss of physical content.

Algebra of bounded observables. We now define $\mathcal{A}$ to be the set of all bounded operators in $\mathcal{H}$: the bounded observables are thus the self-adjoint

† This is the case if the parts $A^{(\pm)}$, whose spectra are positive and negative, respectively, satisfy the condition that $\mathrm{Tr}(\rho A^{(\pm)})$ are both finite.

elements of $\mathscr{A}$. From the definition of $\mathscr{A}$, it follows that this set of operators has an algebraic structure, in that, if λ is a complex number and A, B belong to $\mathscr{A}$, then λA, A^*, $(A+B)$ and AB all belong to $\mathscr{A}$. In view of these properties, $\mathscr{A}$ is termed a **-algebra* (star algebra), the star referring to the fact that $\mathscr{A}$ is closed with respect to Hermitian conjugation. As all the algebras we deal with are *-algebras, we shall generally omit reference to the star. In the present physical context, we call $\mathscr{A}$ the *algebra of bounded observables* of the system. Since $\mathscr{A}$ consists of bounded operators, each of its elements, A, has a norm $\|A\|$ given by the supremum (i.e. least upper bound) of the vector norm $\|Af\|$ as f runs through the normalized vector in $\mathscr{H}$ (cf. Appendix A). Here, $\|A\|$ enjoys the following standard properties of operators norms (cf. Appendix A, eqn (A.8))

$$\left.\begin{aligned} &\|\lambda A\| = |\lambda|\,\|A\|; \qquad \|A^*A\| = \|A\|^2 \\ &\|A+B\| \leqslant \|A\| + \|B\|; \qquad \|AB\| \leqslant \|A\|\,\|B\|. \end{aligned}\right\} \tag{23}$$

The states. The formula (14) may be employed to express the states as functionals on $\mathscr{A}$, i.e.

$$\rho(A) = \mathrm{Tr}(\rho A) \quad \text{for all } A \text{ in } \mathscr{A}. \tag{24}$$

It follows from this formula that, if A, B are elements of $\mathscr{A}$ and λ, μ are complex numbers, then

$$\rho(\lambda A + \mu B) = \lambda\rho(A) + \mu\rho(B) \tag{25}$$

$$\rho(A^*A) \geqslant 0 \tag{26}$$

and

$$\rho(I) = 1. \tag{27}$$

The states are termed *linear*, *positive*, and *normalized* by virtue of the formula (25), (26), and (27), respectively.

Note. In general, the conditions (25)–(27) do not completely characterize the physical states, since it is known that, unless $\mathscr{H}$ is finite-dimensional, there are linear, positive normalized functionals which do not correspond to density matrices according to (24) (cf. Emch 1966).

A further consequence of the definition (24) is that

$$|\rho(A)| \leqslant \|A\| \quad \text{for all } A \text{ in } \mathscr{A}. \tag{28}$$

This implies that ρ has the *continuity* property that, if $\{A_n\}$ is a sequence in $\mathscr{A}$ which converges in norm to A, i.e. if $\|A_n - A\| \to 0$ as $n \to \infty$, then $\rho(A_n) \to \rho(A)$ in this limit.

Mixtures and pure states. It follows from eqn (24) that, if ρ_1 and ρ_2 are states and $0 < \lambda < 1$, then $\lambda\rho_1 + (1-\lambda)\rho_2$ is also a state. The weighted sum $\lambda\rho_1 + (1-\lambda)\rho_2$ is naturally interpreted as a mixture of the states ρ_1 and ρ_2.

Furthermore, the states which cannot be resolved into mixtures of different states are precisely the pure ones, in the sense defined in §2.2.2, since the one-dimensional projectors are the only density matrices that cannot be expressed as weighted sums of other density matrices. Hence we conclude that the earlier definition of pure states is equivalent to the natural one, in which these are characterized by indecomposability into different states.

The dynamics. The dynamics of the system is given, in the Heisenberg scheme, by the formula (18) for the transformations $A \to A_t$ of the observables. Since this transformation is linear, we may extend it to all elements of the algebra $\mathscr{A}$, i.e.

$$A_t = U_t A U_t^* \equiv \alpha_t A \quad \text{for all } A \text{ in } \mathscr{A}. \tag{29}$$

It follows from this equation and the unitarity of U_t that α_t preserves the algebraic structure of $\mathscr{A}$, i.e.

$$\left.\begin{aligned} &\alpha_t(\lambda A) = \lambda \alpha_t A; \qquad \alpha_t(A + B) = \alpha_t A + \alpha_t B; \\ &\alpha_t(AB) = (\alpha_t A)(\alpha_t B); \qquad \alpha_t(A^*) = (\alpha_t A)^*; \\ &\text{and} \quad \|\alpha_t A\| = \|A\|. \end{aligned}\right\} \tag{30}$$

Accordingly, the transformations α_t are termed *automorphisms* of $\mathscr{A}$. By (16) and (29), they possess the group property that

$$\alpha_t \alpha_s = \alpha_{t+s}; \qquad \alpha_0 = I. \tag{31}$$

Thus, the dynamics is represented, in the Heisenberg scheme, by the one-parameter group $\{\alpha_t\}$ of automorphisms of $\mathscr{A}$.

Equivalently, in the Schrödinger scheme, the dynamics is given by the transformations $\rho \to \rho_t$ of the states, where, by (19) and (29),

$$\rho_t(A) = \rho(\alpha_t A). \tag{32}$$

Hence, in this scheme, the dynamics corresponds to the transformations, $\bar{\alpha}_t$, of the states, as defined by the formula

$$\bar{\alpha}_t \rho = \rho_t, \tag{33}$$

and these transformations possess the group property

$$\bar{\alpha}_t \bar{\alpha}_s = \bar{\alpha}_{t+s}. \tag{34}$$

The dynamics is therefore represented, in the Schrödinger scheme, by the one-parameter group $\{\bar{\alpha}_t\}$ of transformations of the state functionals.

2.2.4. *Summary of the model*

We may summarize the model of a finite system as follows. The *bounded observables* are the self-adjoint elements of the algebra $\mathscr{A}$. The *states* are the functionals on $\mathscr{A}$ that correspond to density matrices according to (24).

The *pure states* are those that cannot be expressed as weighted sums, i.e. mixtures, of different states. The *dynamics* is given, in the Heisenberg picture, by the one-parameter group $\{\alpha_t\}$ of automorphisms of $\mathscr{A}$; or equivalently, in the Schrödinger picture, by the group $\{\bar{\alpha}_t\}$ of transformations of the states. Finally, *the unbounded observables* are expressed in terms of the bounded ones by eqn (21).

2.3. Inequivalent representations for an infinite system

As we have just seen, the model of a finite system depends crucially on Von Neumann's theorem of the equivalence of all irreducible representations of its observables. We shall now demonstrate, with a simple example, that, by contrast, an infinite system can support a multitude of inequivalent representations of its observables. Furthermore, we shall show that the different representations which we construct correspond to *macroscopically different* classes of states. The results obtained here will be generalized, in the following Section, to arbitrary infinite systems.

We consider an infinite chain of Pauli spins $\{\boldsymbol{\sigma}_n\}$ located at sites labelled by the integers, $n = 0, \pm 1, \pm 2, \ldots$ Assuming that these spins are measured in units of $\frac{1}{2}\hbar$, a representation of them consists of self-adjoint operators $\{\hat{\boldsymbol{\sigma}}_n = (\hat{\sigma}_{nx}, \hat{\sigma}_{ny}, \hat{\sigma}_{nz})\}$ in a Hilbert space, $\mathscr{H}$, such that

$$\left.\begin{aligned} &[\hat{\sigma}_{nx}, \hat{\sigma}_{ny}] = \mathrm{i}\hat{\sigma}_{nz}; \qquad [\hat{\sigma}_{ny}, \hat{\sigma}_{nz}] = \mathrm{i}\hat{\sigma}_{nz}; \\ &[\hat{\sigma}_{nz}, \hat{\sigma}_{nx}] = \mathrm{i}\hat{\sigma}_{ny}; \quad \hat{\boldsymbol{\sigma}}_n \cdot \hat{\boldsymbol{\sigma}}_n = 3; \\ &\text{and } \hat{\boldsymbol{\sigma}}_m \text{ commutes with } \hat{\boldsymbol{\sigma}}_n \text{ if } m \neq n. \end{aligned}\right\} \tag{35}$$

The first representation we construct is based on configurations of the z-components of the spins, in which all but a finite number of them take the value 1. To be precise, we define a σ_z-configuration to be a sequence s of numbers $\{s_n \mid n = 0, \pm 1, \pm 2, \ldots\}$, each of which takes the values 1 or -1, these being the eigenvalues of each σ_z. We term a configuration s *essentially positive* if all but a finite number of its components, s_n, are equal to $+1$. We assign to each essentially positive configuration s a unit vector $\Psi^{(+)}(s)$, such that $\Psi^{(+)}(s)$ and $\Psi^{(+)}(s')$ are orthogonal if s' is different from s. We then define $\mathscr{H}^{(+)}$ to be the Hilbert space for which the vectors $\{\Psi^{(+)}(s) \mid s$ essentially positive$\}$ form a complete orthonormal basis. In order to obtain a representation of the spins in $\mathscr{H}^{(+)}$, we define a sequence of operators $\{\hat{\boldsymbol{\sigma}}_n^{(+)}\}$ by a natural generalization of the formula (9) for the Pauli representation of a single spin, and then confirm that this sequence $\{\hat{\boldsymbol{\sigma}}_n^{(+)}\}$ does indeed satisfy the conditions (35). Thus, we start by defining $\hat{\boldsymbol{\sigma}}_n^{(+)}$ by the following analogue of eqn (9) for a single spin.

$$\hat{\boldsymbol{\sigma}}_n^{(+)}\Psi^{(+)}(s) = (\Psi^{(+)}(\theta_n s), \mathrm{i}s_n\Psi^{(+)}(\theta_n s), s_n\Psi^{(+)}(s)), \tag{36}$$

where

$$(\theta_n s)_m = \begin{cases} s_m & \text{if } m \neq n. \\ -s_m & \text{if } m = n. \end{cases} \tag{37}$$

It is now a simple matter to check that the components of the $\hat{\boldsymbol{\sigma}}_n^{(+)}$'s are self-adjoint and satisfy the conditions (35). Hence $\{\hat{\boldsymbol{\sigma}}_n^{(+)}\}$ is a representation of the spins in $\mathscr{H}^{(+)}$. Furthermore, it is irreducible, since any essentially positive configuration s' may be obtained from any other, s, by reversal of a finite number of s_n's and therefore, by (36), there is a corresponding product $\hat{\sigma}_{n_1x}^{(+)} \ldots \hat{\sigma}_{n_kx}^{(+)}$ which leads from $\Psi^{(+)}(s)$ to $\Psi^{(+)}(s')$.

We may similarly construct another representation, based on configurations, $s = \{s_n\}$, for which all but a finite number of the s_n's are equal to -1. Thus, we assign a unit vector $\Psi^{(-)}(s)$ to each such configuration s, with $\Psi^{(-)}(s)$ and $\Psi^{(-)}(s')$ orthogonal if $s \neq s'$; and we define $\mathscr{H}^{(-)}$ to be the Hilbert space for which the resultant set $\{\Psi^{(-)}(s)\}$ of orthonormal vectors forms a basis. We then define the operators $\{\hat{\boldsymbol{\sigma}}_n^{(-)}\}$ in $\mathscr{H}^{(-)}$ by analogy with (36), i.e.

$$\hat{\boldsymbol{\sigma}}_n^{(-)}\Psi^{(-)}(s) = (\Psi^{(-)}(\theta_n s), \mathrm{i}s_n\Psi^{(-)}(\theta_n s), s_n\Psi^{(-)}(s)), \tag{38}$$

and we verify that these operators constitute an irreducible representation of the spins in $\mathscr{H}^{(-)}$.

Thus, we have constructed two representations of the spins, the first by the operators $\{\hat{\boldsymbol{\sigma}}_n^{(+)}\}$ in $\mathscr{H}^{(+)}$ and the second by $\{\hat{\boldsymbol{\sigma}}_n^{(-)}\}$ in $\mathscr{H}^{(-)}$. It is rather obvious that these representations are physically inequivalent, since they are based on completely different spin configurations. We shall now demonstrate their inequivalence by showing that the spin density of the entire system takes the value $\mathbf{k}$, the unit vector along $0z$, for the states given by density matrices in the first representation; and $-\mathbf{k}$ for those corresponding to density matrixes in the second one. Thus, defining

$$\hat{\mathbf{m}}_N^{(+)} = \frac{1}{2N+1}\sum_{-N}^{N} \hat{\boldsymbol{\sigma}}_n^{(+)}, \tag{39}$$

the $\hat{\boldsymbol{\sigma}}^{(+)}$-representation of the mean density of the spins between the sites $-N$ and N, we deduce from eqns (36) and (39) that

$$(\Psi^{(+)}(s), \hat{\mathbf{m}}_N^{(+)}\Psi^{(+)}(s)) = \frac{1}{2N+1}\sum_{-N}^{N} \hat{\boldsymbol{\sigma}}_n^{(+)}, \tag{40}$$

Hence, as all but a finite number of the s_n's take the value $+1$,

$$(\Psi^{(+)}(s), \hat{\mathbf{m}}_N^{(+)}\Psi^{(+)}(s)) \to \mathbf{k} \quad \text{as} \quad N \to \infty. \tag{41}$$

Similarly, by (36) and (39),

$$(\Psi^{(+)}(s), \hat{\mathbf{m}}_N^{(+)}\Psi^{(+)}(s')) \to 0 \quad \text{as} \quad N \to \infty \text{ if } s' \neq s. \tag{42}$$

Since any normalized vector $\Psi^{(+)}$ in $\mathscr{H}^{(+)}$ is of the form $\sum c(s)\Psi^{(+)}(s)$, with $\sum |c(s)|^2 = 1$, it follows from (41) and (42) that†

$$(\Psi^{(+)}, \hat{\mathbf{m}}_N^{(+)}\Psi^{(+)}) \to \mathbf{k} \quad \text{as} \quad N \to \infty. \tag{43}$$

Hence, for any density matrix $\rho^{(+)}$ in $\mathscr{H}^{(+)}$,

$$\mathrm{Tr}(\hat{\rho}^{(+)}\hat{m}_N^{(+)}) \to \mathbf{k} \quad \text{as} \quad N \to \infty. \tag{44}$$

This means that the polarization (spin density) of the entire system is **k** for all states given by density matrices in the $\hat{\sigma}^{(+)}$-representation. Similarly, the polarization is $-\mathbf{k}$ for all states given by density matrices in the $\hat{\sigma}^{(-)}$-representation. Since the polarization is a macroscopic, i.e. global, variable, it follows that the two representations are *macroscopically inequivalent*. This also implies that they are not unitarily equivalent: for were they so, then it would follow that the expectation values of the spin density were the same for the states associated with the two representations, and we have just seen that this is not the case.

The method employed here may be adapted for the construction of an enormous variety of inequivalent representations of the spins. For example, one may use the same method to obtain a representation based on configurations in which all but a finite number of the spins are orientated along a direction given by some unit vector **u**, instead of **k**. The polarization of the system for the states associated with this representation would then be **u**. Since **u** may be arbitrarily chosen, it follows that the system supports an infinity of inequivalent representations, with macroscopically different properties. Furthermore, it is not difficult to construct other representations, with macroscopic properties different from those we have considered so far. For example, one could base a representation on 'antiferromagnetic' configurations, in which the spins, $\boldsymbol{\sigma}_n$, at all but a finite number of sites, are oriented along the direction **u** or $-\mathbf{u}$, according to whether n is odd or even.

Discussion: *representations and phases*. In order to relate the representations we have constructed to the interactions in the system, we consider the case of a Heisenberg model with formal Hamiltonian

$$H = -J \sum_n \boldsymbol{\sigma}_n \cdot \boldsymbol{\sigma}_{n+1}, \quad \text{with} \quad J > 0. \tag{45}$$

The state in which the z-components of the spins are all aligned parallel to $0z$ is then a ground state. It corresponds to the vector $\Psi^{(+)}(\bar{s})$ in the $\hat{\sigma}^{(+)}$-representation, where $\bar{s}$ is the configuration whose components, $\bar{s}_n$, are

† To be precise, (43) follows directly from (41) and (42) when $\Psi^{(+)}$ is a *finite* linear combination $\sum c(s)\Psi^{(+)}(s)$ of the basis vectors $\Psi^{(+)}(s)$. Since such linear combinations are dense in $\mathscr{H}^{(+)}$, (43) may be extended to all normalized vectors $\Psi^{(+)}$ in $\mathscr{H}^{(+)}$ by virtue of the fact that, by (39), the norms of $\hat{m}_{Nx}^{(+)}$, $\hat{m}_{Ny}^{(+)}$ and $\hat{m}_{Nz}^{(+)}$ cannot exceed unity.

all equal to +1. Furthermore, as any of the basis vectors $\Psi^{(+)}(s)$ in $\mathscr{H}^{(+)}$ is obtained from the ground state $\Psi^{(+)}(\bar{s})$ by reversal of a finite number of spins, it follows that the full set of states given by the density matrices in $\mathscr{H}^{(+)}$ is generated by localized modifications of $\Psi^{(+)}(\bar{s})$, corresponding to such spin reversals. This set of states may therefore be taken to comprise a *phase*, generated by localized modifications of the ground state, $\Psi^{(+)}(\bar{s})$. It is then a phase of the system at zero temperature and, as shown above, all the states in this phase have the same polarization, $\mathbf{k}$.

Likewise, for any arbitrary direction $\mathbf{u}$, the system supports a phase based on the configuration in which all spins are aligned parallel to $\mathbf{u}$, at absolute zero. The states in this phase then all have polarization $\mathbf{u}$. Thus we have a set of different phases of the system corresponding to different representations of the observables, for the same temperature (0 °K in this case). Furthermore, although the basic Hamiltonian H is invariant under rotations of the spin vectors, each of these phases has a polarization along a definite direction. Thus we have a *spontaneous symmetry breakdown*, as each phase lacks the rotational symmetry of the interactions in the system. This is a situation which typifies a class of phase transitions (cf. Landau and Lifshitz 1959, Ch. 14). We emphasize here that this situation could not be covered by a model of a finite system, since that would admit only one representation of its observables and therefore would not present the phase structure we have just described.

In general, we shall refer to the phenomenon whereby a system exhibits more than one phase at the same temperature (zero in this case) as *macroscopic degeneracy*.

2.4. General model of an infinite system

Suppose now that Σ is an arbitrary infinitely extended system of particles in a space X, which we take, for simplicity, to be either a Euclidean continuum or a simple cubic lattice. The model of Σ is based on the assumption† that its observables in any bounded region Λ are those of a finite system, $\Sigma(\Lambda)$, of particles of the given species confined to Λ. The local observables for Σ are taken to comprise those of all the bounded regions in X. The states of Σ are assumed to be functionals of the observables, which reduce, in each bounded region Λ, to states of the finite system $\Sigma(\Lambda)$. The dynamics of Σ is defined so that the time-dependent expectation values of its observables are suitable limits of those of a corresponding finite system. Thus the model of Σ is based on a natural generalization of quantum theory from finite to infinite systems.

We shall present the construction and basic features of this model as follows. In §2.4.1 and 2.4.2, we shall formulate the bounded observables

† The ideas concerning the local structure assumed here were first proposed by Haag and Kastler (1964).

and states, respectively, of Σ. In §2.4.3, we shall specify a general connection between the states of Σ and the representations of its observables. In §2.4.4, we shall show that a representation corresponds to a class of macroscopically equivalent states, as in the particular model of §2.3. In §2.4.5, we shall formulate the dynamics of Σ as a limiting form of that of a corresponding finite system. In §2.4.6, we shall extend the model so as to include the unbounded observables. In §2.4.7, we shall summarize the structure of the model.

2.4.1. The local observables

We formulate the local observables of the system according to the following scheme. For each bounded, open† spatial region Λ, we construct a Hilbert space, $\mathscr{H}(\Lambda)$, and an algebra, $\mathscr{A}(\Lambda)$, of bounded operators in $\mathscr{H}(\Lambda)$ in such a way that $\mathscr{A}(\Lambda)$ is the algebra of bounded observables for the finite system, $\Sigma(\Lambda)$, of particles of the given species in Λ. The explicit details of this construction will be provided at the end of this subsection. As regards the unbounded observables for the region Λ, say, these may be expressed as limits of the bounded ones, as in the case of a finite system (cf. eqn (21)). We shall denote by $H(\Lambda)$ the energy observable for this region. This will correspond to the Hamiltonian for $\Sigma(\Lambda)$.

The family of algebras $\{\mathscr{A}(\Lambda)\}$ is constructed so as to satisfy the following natural conditions.

($\mathscr{A}1$) If $\Lambda \subset \Lambda'$, then $\mathscr{A}(\Lambda) \subset \mathscr{A}(\Lambda')$; and the restriction to $\mathscr{A}(\Lambda)$ of the $\mathscr{A}(\Lambda')$-norm is identical with the $\mathscr{A}(\Lambda)$-norm. Thus, observables for the region Λ are also observables for any region containing Λ. $\mathscr{A}(\Lambda)$ is said to be *isotonic* with respect to Λ, by virtue of ($\mathscr{A}1$). It follows immediately from this property that if Λ and Λ' are arbitrary bounded regions, then $\mathscr{A}(\Lambda)$ and $\mathscr{A}(\Lambda')$ are subalgebras of $\mathscr{A}(\Lambda \cup \Lambda')$. This implies that elements of $\mathscr{A}(\Lambda)$ and $\mathscr{A}(\Lambda')$ may be added or multiplied together, and thus permits the statement of the following condition.

($\mathscr{A}2$) If Λ and Λ' are disjoint, then the elements of $\mathscr{A}(\Lambda)$ commute with those of $\mathscr{A}(\Lambda')$.

We refer to ($\mathscr{A}2$) as the local *commutativity* condition. it is needed to ensure that observables in disjoint regions may be measured simultaneously.

In order to obtain a comprehensive description of the local observables of Σ, we note again that, by ($\mathscr{A}1$), the algebras $\mathscr{A}(\Lambda)$ and $\mathscr{A}(\Lambda')$ for arbitrary bounded regions Λ and Λ', are subalgebras of $\mathscr{A}(\Lambda \cup \Lambda')$. Thus, we may form the union $\mathscr{A}(\Lambda) \cup \mathscr{A}(\Lambda')$, consisting of those elements of $\mathscr{A}(\Lambda \cup \Lambda')$ that belong to $\mathscr{A}(\Lambda)$ or $\mathscr{A}(\Lambda')$. Hence, by induction, we may form the union, $\mathscr{A} = \bigcup \mathscr{A}(\Lambda)$, of all the local algebras $\mathscr{A}(\Lambda)$. It follows from this

† In the case where X is a lattice, the bounded open regions are those that consist of finite sets of sites. In the case where X is a continuum, they are the interiors of the bounded regions.

definition that $\mathscr{A}$ inherits an algebraic structure from the $\mathscr{A}(\Lambda)$'s, whereby it is closed with respect to binary multiplication and addition, multiplication by complex numbers and Hermitian conjugation. It is therefore a *-algebra (cf. §2.2.3). In the present physical context, we term $\mathscr{A}$ the *algebra of bounded local observables*† of Σ, since its self-adjoint elements are the bounded observables for the various finite regions. It follows from ($\mathscr{A}$1) and the definition of $\mathscr{A}$ that this algebra is equipped with a norm which inherits from the $\mathscr{A}(\Lambda)$'s the following standard operator-norm properties (cf. (23)).

$$\left.\begin{aligned} &\|\lambda A\| = |\lambda|\,\|A\|; \qquad \|A^*A\| = \|A\|^2; \\ &\|A+B\| \leqslant \|A\| + \|B\|; \qquad \|AB\| \leqslant \|A\|\,\|B\|. \end{aligned}\right\} \tag{46}$$

Note. According to the above definition, $\mathscr{A}$, unlike the $\mathscr{A}(\Lambda)$'s, is not an algebra of operators in a Hilbert space, but rather a set of elements which inherits an algebraic structure from the latter algebras. Of course, as we shall see in §2.4.3, $\mathscr{A}$ does admit Hilbert space representations, but since there are a multitude of inequivalent ones, as in the model of §2.3, one would lose generality by limiting the description of the observables to any particular representation.

As in §2.2, *automorphisms* of $\mathscr{A}$ are defined to be transformations, γ of this algebra which preserve its algebraic structure, i.e.

$$\left.\begin{aligned} &\gamma(\lambda A) = \lambda\gamma A; \qquad \gamma(A+B) = \gamma A + \gamma B; \qquad \gamma(AB) = (\gamma A)(\gamma B); \\ &\gamma(A^*) = (\gamma A)^*; \quad \text{and} \quad \|\gamma A\| = \|A\|. \end{aligned}\right\} \tag{47}$$

We define space-translates of the local observables in terms of automorphisms of $\mathscr{A}$. Thus, denoting the points of the space X by x, we stipulate that the local algebras $\{\mathscr{A}(\Lambda)\}$ are so constructed that:

($\mathscr{A}$3) $\mathscr{A}$ is equipped with a group $\{\gamma(x) \mid \gamma(x_1)\gamma(x_2) = \gamma(x_1 + x_2)\}$ of automorphisms, which represent the space-translations $\{x\}$ and satisfy the covariance condition

$$\gamma(x)\mathscr{A}(\Lambda) = \mathscr{A}(\Lambda + x), \tag{48}$$

where $(\Lambda + x)$ is the space-translate of Λ by the displacement x. We shall generally denote the space-translate of an observable A by $A(x)$, i.e.

$$\gamma(x)A \equiv A(x). \tag{49}$$

Explicit constructions. We shall now provide explicit constructions of the local algebras of both lattice and continuous systems in accordance with the requirements of ($\mathscr{A}$1–3).

† We remark here that it is standard, in the current literature, to take the algebra of local observables to be that obtained by adding to $\mathscr{A}$ the limits of its norm-convergent Cauchy sequences. However, we find it unnecessary to look beyond $\mathscr{A}$ for our description of the local bounded observables.

***Example* 1. *Lattice system*.** Suppose that Σ is a system of identical atoms, or spins, located on the sites of a simple cubic lattice, X: the atoms are then distinguishable by the sites at which they are localized. We assume that the observables and pure states of each atom, when isolated, are given by the self-adjoint operators and normalized vectors, respectively, in a finite-dimensional Hilbert space $\mathscr{H}_0$. Correspondingly, we represent the observables of the atom at the site x by the self-adjoint operators in a copy, $\mathscr{H}_x$, of $\mathscr{H}_0$. For each bounded region, i.e. each finite set of sites, Λ, we then define $\mathscr{H}(\Lambda)$ to be the tensor product† $\bigotimes_{x\in\Lambda}\mathscr{H}_x$, and $\mathscr{A}(\Lambda)$ to be the set of all operators in $\mathscr{H}(\Lambda)$ – these operators are all bounded, since $\mathscr{H}_0$, and therefore $\mathscr{H}(\Lambda)$, is finite-dimensional.

It follows from our definition of $\mathscr{H}(\Lambda)$ that, if $\Lambda\subset\Lambda'$, then $\mathscr{H}(\Lambda')=\mathscr{H}(\Lambda)\otimes\mathscr{H}(\Lambda'\backslash\Lambda)$, where $\Lambda'\backslash\Lambda$ is the complement of Λ in Λ'. Since the algebras $\mathscr{A}(\Lambda)$, in $\mathscr{H}(\Lambda)$, and $\mathscr{A}(\Lambda)\otimes I_{\Lambda'\backslash\Lambda}$, in $\mathscr{H}(\Lambda')$, have identical structures, we identify an element, A, of $\mathscr{A}(\Lambda)$ with the corresponding one, $A\otimes I_{\Lambda'\backslash\Lambda}$, of $\mathscr{A}(\Lambda')$: this amounts to saying that these operators represent the same observable.

It is now a simple matter to check that the isotony property ($\mathscr{A}1$) and the local commutativity ($\mathscr{A}2$) are consequences of our definition of the local algebras $\{\mathscr{A}(\Lambda)\}$ and the identification of $\mathscr{A}(\Lambda)$ with $\mathscr{A}(\Lambda)\otimes I_{\Lambda'\backslash\Lambda}$, which we have just described. We now define the algebra, $\mathscr{A}$, of bounded local observables as the union of the local algebras $\{\mathscr{A}(\Lambda)\}$. In order to define the space-translational automorphisms of $\mathscr{A}$ we note that, as each $\mathscr{H}_x$ is a copy of $\mathscr{H}_0$, there is a one-to-one correspondence between the algebra $\mathscr{A}_x$ of observables for the site x and the algebra, $\mathscr{A}_0$, of operators in $\mathscr{H}_0$. Thus, denoting by a_x the element of $\mathscr{A}_x$ corresponding to a, of $\mathscr{A}_0$, it follows from the definition of $\mathscr{A}$ that this algebra consists of linear combinations of terms $a_{x_1}^{(1)}\ldots a_{x_k}^{(k)}$, where $x_1,\ldots x_k$ and $a^{(1)},\ldots,a^{(k)}$ run through X and $\mathscr{A}_0$, respectively. We define $\gamma(x)$ to be the linear transformation of $\mathscr{A}$, corresponding to space translation by x, by the formula‡

$$\gamma(x)(a_{x_1}^{(1)}\ldots a_{x_k}^{(k)})=a_{x_1+x}^{(1)}\ldots a_{x_k+x}^{(k)}. \tag{50}$$

From this definition it follows that $\gamma(x)$ is an automorphism of $\mathscr{A}$, and that the set $\{\gamma(x)\}$ of these automorphisms satisfies ($\mathscr{A}3$).

Since the atoms are assumed to be fixed at the lattice sites, we take the local Hamiltonian $H(\Lambda)$ to be the total potential energy between the particles in Λ. In general, this energy may arise not only from two-body but also from many-body interactions of various orders. We represent the r-body interaction between the atoms at $x_1,\ldots,x_r$ by a potential energy function $v_r(x_1,\ldots,x_r)$ which we take to be a Hermitian element of the local algebra $\mathscr{A}(\{x_1,\ldots,x_r\})$. We then define the local Hamiltonian $H(\Lambda)$ by the

† For a definition of tensor products, see Appendix A.
‡ Here we identify $a_{x_1}^{(1)}\ldots a_{x_k}^{(k)}$ with the tensor product $a_{x_1}^{(1)}\otimes a_{x_2}^{(2)}\otimes\ldots\otimes a_{x_k}^{(k)}$.

formula

$$H(\Lambda) = \sum_r \sum_{\substack{x_1 \ldots x_r \\ \in \Lambda}} v_r(x_1, \ldots, x_r). \tag{51}$$

***Example* 2. *Continuous systems*.** Suppose, for simplicity, that Σ is a system of bosons or fermions, of one species, in a Euclidean space X. We formulate the observables of Σ in second quantization,† thereby permitting the number of particles in any bounded region to be variable. Thus, we describe the observables in terms of a quantized field ψ, which satisfies the canonical commutation or anticommutation relations, according to whether the system consists of bosons or fermions, i.e. formally,

$$[\psi(x), \psi^*(x')]_\mp = \delta(x - x'); \qquad [\psi(x), \psi(x')]_\mp = 0, \tag{52}$$

where

$$[A, B]_\mp = AB \mp BA, \tag{53}$$

and where spin has been ignored, for simplicity. In order to give the formal relations (52) a precise mathematical meaning, we need to 'smear out' the field ψ by integrating it against suitable test-functions and define the resultant smeared field $\psi(f) = \int \mathrm{d}x \psi(x) f(x)$ as an operator in a Hilbert space. Thus, we define both the Fock–Hilbert space $\mathscr{H}_\mathrm{F}$ and the operators $\psi(f)$ in $\mathscr{H}_\mathrm{F}$ by the following standard specifications (cf. Fock 1932, Cook 1953).

(1) For each square-integrable function $f(x)$, there is an operator $\psi(f)$ in $\mathscr{H}_\mathrm{F}$ such that

$$[\psi(f), \psi(g)^*]_\mp = \int \mathrm{d}x \bar{g}(x) f(x); \qquad [\psi(f), \psi(g)]_\mp = 0. \tag{54}$$

These relations may be formally derived from (52) by putting $\psi(f) = \int \mathrm{d}x \psi(x) f(x)$.

(2) There is a vector Ψ_F in $\mathscr{H}_\mathrm{F}$ such that

$$\psi(f)\Psi_\mathrm{F} = 0 \tag{55}$$

for all the square-integrable functions f.

(3) The vectors obtained by application to Ψ_F of all the polynomials in the $\psi(f)^*$'s form a dense subset of $\mathscr{H}_\mathrm{F}$, i.e. this space is generated by applying these polynomials to Ψ_F. Thus, in a standard terminology, Ψ_F is a vacuum vector, the $\psi(f)^*$'s and $\psi(f)$'s are creation and annihilation operators, respectively, and the various many-particle states are obtained by applying combinations of creation operators to the vacuum. We represent space

† A useful introduction to the theory of second quantization is provided in Chapter 6 of Schweber's book (1961). Rigorous treatments are given by Fock (1932) and Cook (1953).

translations by the group of unitary operators $\{V(x)\}$ which leave Ψ_F invariant and implement the formal relation $V(x)\psi(x')V(x)^* = \psi(x+x')$. To be precise, $V(x)$ is defined by the formula

$$V(x)\Psi_F = \Psi_F; \qquad V(x)\psi(f_1)^* \ldots \psi(f_k)^*\Psi_F = \psi(f_{1,x})^* \ldots \psi(f_{k,x})^*\Psi_F \quad (56)$$

where

$$f_x(x') = f(x'-x). \quad (57)$$

Likewise, we represent gauge transformations by the unitary group $\{W(\alpha) \mid \alpha \text{ real}\}$ which leave Ψ_F invariant and implement the formal relation $W(\alpha)\psi(x)W(\alpha)^* = \psi(x)e^{i\alpha}$. The precise definition of $W(\alpha)$ is given by the formula

$$W(\alpha)\Psi_F = \Psi_F; \qquad W(\alpha)\psi(f_1)^* \ldots \psi(f_k)^*\Psi_F = e^{-ik\alpha}\psi(f_1)^* \ldots \psi(f_k)^*\Psi_F. \quad (58)$$

Turning now to the local description, we define the Hilbert space $\mathscr{H}(\Lambda)$, for an open bounded region Λ, to be the subspace of $\mathscr{H}_F$ generated by applying the polynomials in those $\psi(f)^*$'s, for which f vanishes outside Λ, to the vacuum vector Ψ_F. It follows from this definition and eqns (56) and (58) that $V(x)$ transforms $\mathscr{H}(\Lambda)$ into $\mathscr{H}(\Lambda+x)$ and that $W(\alpha)$ maps $\mathscr{H}(\Lambda)$ onto itself. Thus, the restriction of $W(\alpha)$ to $\mathscr{H}(\Lambda)$ is a unitary transformation of this space.

We define $\mathscr{A}(\Lambda)$, the algebra of bounded observables for the region Λ, to be the set of bounded operators in $\mathscr{H}(\Lambda)$ which commute with the gauge transformations $\{W(\alpha)\}$. These are just the bounded operators in $\mathscr{H}(\Lambda)$ that commute with the local number operator $N(\Lambda) = \int_\Lambda dx\,\psi^*(x)\psi(x)$, and the n-particle projection of $\mathscr{A}(\Lambda)$ is equivalent to the algebra of bounded observables for an n-particle system of the given species, as described in first quantization.

It follows immediately from the definition of $\mathscr{A}(\Lambda)$ that the isotony condition ($\mathscr{A}1$) is satisfied. It may also be checked that, in view of the canonical commutation or anticommutation relations (54) and the gauge invariance of the $\mathscr{A}(\Lambda)$'s, these algebras also satisfy the local commutativity condition ($\mathscr{A}2$).†

As previously, we define the algebra, $\mathscr{A}$, of bounded local observables of the system to be the union of the local algebras $\{\mathscr{A}(\Lambda)\}$. We define the space-translational automorphisms, $\gamma(x)$, of this algebra by the formula

$$\gamma(x)A = V(x)AV(x)^*, \quad (59)$$

and we verify that, by (56), (59), and our definition of $\mathscr{A}(\Lambda)$, these automorphisms satisfy ($\mathscr{A}3$).

† This would not be true, in the case of fermions, if we took $\mathscr{A}(\Lambda)$ to be the set of all bounded operators in $\mathscr{H}(\Lambda)$. For then $\mathscr{A}(\Lambda)$ would include the smeared fields $\psi(f)$, localized in Λ; and such fermion fields for disjoint regions Λ and Λ' do not commute but, instead, anticommute.

The unbounded observables of the region Λ are the unbounded self-adjoint operators in $\mathscr{H}(\Lambda)$ that commute with the gauge transformations, $W(\alpha)$, or equivalently with the local number operator $N(\Lambda)=\int_{\Lambda}\mathrm{d}x\times \psi^*(x)\psi(x)$. These observables generally include the local energy observable, or Hamiltonian, $H(\Lambda)$. In the case where Σ consists of particles of mass m, which interact via conservative two-body forces, $H(\Lambda)$ takes the form

$$H(\Lambda)=\frac{\hbar^2}{2m}\int_{\Lambda}\mathrm{d}x\nabla\psi^*(x)\,.\,\nabla\psi(x)$$
$$+\tfrac{1}{2}\int_{\Lambda}\mathrm{d}x\int_{\Lambda}\mathrm{d}x'\psi^*(x)\psi^*(x')V(x,x')\psi(x)\psi(x'), \tag{60}$$

where $V(x,x')$ is the potential energy of interaction between particles at x and x'.

Note. There are two ways of looking at the algebra, $\mathscr{A}$, of bounded local observables of this system. The first is to regard it as an abstract algebra, whose structure is inherited from the local operator algebras $\{\mathscr{A}(\Lambda)\}$. The second is to regard it as an algebra, $\mathscr{A}_{\mathrm{F}}$, of operators in $\mathscr{H}_{\mathrm{F}}$. For reasons explained in the Note following eqn (46), we adopt the first viewpoint, and regard $\mathscr{A}_{\mathrm{F}}$ as just a particular representation of $\mathscr{A}$.

2.4.2. *The states*

The states of Σ are assumed to be functionals of the local observables which reduce, in each bounded region Λ, to states of the finite system $\Sigma(\Lambda)$. Thus, we define a state, ρ, of Σ to be a functional on $\mathscr{A}$ which corresponds to a family of density matrices $\{\rho_{\Lambda}\}$ in the respective Hilbert spaces $\{\mathscr{H}(\Lambda)\}$ according to the formula

$$\rho(A)=\mathrm{Tr}_{\Lambda}(\rho_{\Lambda}A)\quad \text{for } A \text{ in } \mathscr{A}(\Lambda), \tag{61}$$

where Tr_{Λ} denotes the Trace over $\mathscr{H}(\Lambda)$. This formula implies that the ρ_{Λ}'s must satisfy the consistency condition

$$\mathrm{Tr}_{\Lambda}(\rho_{\Lambda}A)=\mathrm{Tr}_{\Lambda'}(\rho_{\Lambda'}A)\qquad \text{if } A \text{ belongs to } \mathscr{A}(\Lambda) \text{ and } \Lambda\subset\Lambda'. \tag{62}$$

In the case where Σ is formulated in second quantization, as in Example 2 of §2.4.1, we assume that the ρ_{Λ}'s, like the local algebras, $\mathscr{A}(\Lambda)$, are gauge invariant.

It follows from eqn (61) that the states of Σ, like those of a finite system, possess the properties of linearity, positivity and normalization given by the formulae

$$\rho(\lambda A+\mu B)=\lambda\rho(A)+\mu\rho(B) \tag{63}$$
$$\rho(A^*A)\geqslant 0 \tag{64}$$

and

$$\rho(I) = 1. \tag{65}$$

It also follows from (61) that

$$|\rho(A)\| \leqslant \|A\|. \tag{66}$$

A theorem due to Dell'Antonio, Doplicher, and Ruelle (1966) reveals that the physical states, as defined by (61), are just those linear, positive, normalized functionals on $\mathscr{A}$ for which there is zero probability of finding an infinite number of particles in some bounded spatial region. We refer to the functionals possessing this latter property as *locally finite*.† Thus, the formula (61) defines the physical states as the linear, positive, normalized, locally finite functionals on $\mathscr{A}$.

It follows from (61) that, if ρ_1 and ρ_2 are states of Σ and $0<\lambda<1$, then $\lambda\rho_1 + (1-\lambda)\rho_2$ is also a state. In other words, mixtures of states are states. We define the *pure states* to be those that cannot be resolved into mixtures of different states:‡ this definition is analogous to that of pure states of finite systems, as may be seen from the discussion in §2.1.4.

An important class of states, which includes the pure ones, consists of those which have *short-range correlations* in the sense specified by Lanford and Ruelle (1969) as follows: given a bounded local observable B and a positive number ε, there exists a bounded region Λ such that, if A is any bounded observable for the complementary region $X\backslash\Lambda$, then

$$|\rho(AB) - \rho(A)\rho(B)| < \varepsilon\, \|A\|. \tag{67}$$

This is a stronger condition than that given by

$$\rho(AB(x)) - \rho(A)\rho(B(x)) \to 0 \text{ as } |x| \to \infty \tag{68}$$

for all local observables A, B. Further, it has been shown by Ruelle (1969*b*) that the states with short-range correlations are 'building-blocks' for the states of Σ in that any state of the system can be resolved into these. This implies that the pure states have short-range correlations, since they cannot be resolved further.

2.4.3. *States, representations, and islands*

An intimate connection between the states of Σ and the representations of its observables follows from a theorem due to Gelfand and Naimark (1943) and Segal (1947). Here, in a usual way, a representation is a mapping, $A \to \hat{A}$, of $\mathscr{A}$ into operators in a Hilbert space $\mathscr{H}$, which preserves the

† The term 'locally normal' is more commonly used in the current literature to describe functionals with this property.

‡ Thus, in mathematical terms, the pure states are the extremal elements of the convex set of all the states.

algebraic structure of $\mathscr{A}$, i.e.

$$\widehat{(\lambda A)} = \lambda \hat{A}; \qquad \widehat{(A+B)} = \hat{A} + \hat{B}; \qquad \widehat{(AB)} = \hat{A}\hat{B}; \qquad \widehat{(A^*)} = (\hat{A})^*. \quad (69)$$

Thus the set of operators, $\hat{A}$, representing the elements of $\mathscr{A}$, form a *-algebra, which we denote by $\hat{\mathscr{A}}$.

The Gelfand–Naimark–Segal (GNS) theorem tells us that for each state, ρ, there is a representation $\hat{\mathscr{A}}$, of $\mathscr{A}$, which is determined up to unitary equivalence, by the following conditions.

(1) There is a normalized vector Ψ in $\mathscr{H}$, such that

$$\rho(A) = (\Psi, \hat{A}\Psi) \quad \text{for all } A \text{ in } \mathscr{A}. \tag{70}$$

(2) The space $\mathscr{H}$ is generated by applying the elements of $\hat{\mathscr{A}}$ to Ψ, i.e. $\hat{\mathscr{A}}\Psi$ is dense in $\mathscr{H}$. The vector Ψ is termed *cyclic* by virtue of this property.

$\hat{\mathscr{A}}$ is usually referred to as the GNS representation of $\mathscr{A}$, and $\mathscr{H}$ as the GNS space, for the state ρ. The essential features of the construction of the GNS representation may be described as follows. For each A in $\mathscr{A}$, one constructs a vector Ψ_A in such a way that Ψ_A is linear in A and the inner product between Ψ_A and Ψ_B is $(\Psi_A, \Psi_B) \equiv \rho(A^*B)$. The space $\mathscr{H}$ is obtained by adding the limit points† to the set of vectors $\{\Psi_A\}$, and the representation of $\mathscr{A}$ in $\mathscr{H}$ is given by the formula $\hat{A}\Psi_B = \Psi_{AB}$. Thus, defining $\Psi = \Psi_I/\|\Psi_I\|$, one checks that the conditions (1) and (2) are fulfilled.

Comments. (a) By eqn (70), ρ corresponds to a vector Ψ in $\mathscr{H}$ irrespective of whether or not ρ is a pure state. However, the GNS representation of $\mathscr{A}$ for this state is irreducible if and only if ρ is pure (cf. Emch 1972, p. 87).

(b) The GNS theorem is applicable not only in the present physical context but also quite generally when $\mathscr{A}$ is an arbitrary *-algebra and ρ a linear, positive, normalized functional on $\mathscr{A}$.

Example. The representations $\hat{\sigma}^{(\pm)}$ of the observables of the spin system which were explicitly constructed in §2.3, are those of GNS for the states in which the spins are aligned along $\pm 0z$, respectively.

The physical significance of the GNS representation is that the functionals, ρ', on $\mathscr{A}$, which correspond to the density matrices, $\hat{\rho}'$, in $\mathscr{H}$, according to the formula

$$\rho'(A) = \mathrm{Tr}(\hat{\rho}'\hat{A}), \tag{71}$$

are states generated by localized modifications of ρ. This may be seen from the following argument. Since $\mathscr{A}$ is the algebra of bounded local observ-

† Thus, if $\{A_n\}$ is a sequence of elements of $\mathscr{A}$ for which the vectors $\{\Psi_{A_n}\}$ possess the Cauchy property that $\lim_{m\to\infty} \lim_{n\to\infty} \|\Psi_{A_m} - \Psi_{A_n}\| = 0$, then we add to this sequence a limit Φ, given by the condition that $\|\Psi_{A_n} - \Phi\| \to 0$ as $n \to \infty$.

ables, the operators, $\hat{A}$, which represent it in $\mathcal{H}$, correspond to localized operations. Hence, in view of the cyclicity of Ψ (cf. condition (2) above), $\mathcal{H}$ is generated by applying localized operations to Ψ. Therefore, as this vector corresponds to the state ρ, by (70), it follows that the functionals, ρ_Φ, corresponding to the normalized vectors, Φ, in $\mathcal{H}$ according to the formula

$$\rho_\Phi(A) = (\Phi, \hat{A}, \Phi), \tag{72}$$

are generated by localized modifications of ρ. The same is true for the functionals ρ', corresponding to the density matrices in $\mathcal{H}$, since these are weighted sum of the ρ_Φ's. Moreover, these functionals ρ' have been shown by the author (Sewell 1970) to be physical states: thus, not only are these functionals linear, positive and normalized, as follows from (61), but they are also locally finite, as one might anticipate from the fact that they are generated by localized modifications of ρ.

We denote by $\mathcal{I}(\rho)$ the set of states, ρ', corresponding to the density matrices, $\hat{\rho}'$, in $\mathcal{H}$ according to (71). This set then includes the vector states, ρ_Φ, as those for which $\hat{\rho}'$ is a one-dimensional projector. Since the localized modifications of ρ which generate $\mathcal{I}(\rho)$ do not lead outside this set of states, we refer to $\mathcal{I}(\rho)$ as the *island of* ρ. We shall presently see that, in the cases where ρ has short-range correlations, the states in this island are macroscopically equivalent to one another.

2.4.4. *Macroscopic description*

We formulate the macroscopic description of the system in terms of global intensive variables. The simplest of these are the infinite volume limits of the observables, $\tilde{A}(\Lambda)$, given by the space-averages of bounded local ones, A, over bounded regions, Λ, i.e.

$$\tilde{A}(\Lambda) = \left.\begin{array}{ll} |\Lambda|^{-1} \sum_{x\Lambda} \gamma(x)A & \text{for a lattice system} \\ |\Lambda|^{-1} \int_\Lambda \mathrm{d}x \gamma(x) A & \text{for a continuous system} \end{array}\right\} \tag{73}$$

where $|\Lambda|$ is the number of sites in Λ in the first case and the volume of Λ in the second one. Assuming that $\rho(\tilde{A}(\Lambda))$ converges as Λ increases to infinity over an increasing sequence of bounded regions, we denote its limit by $\rho(\tilde{A})$, i.e.

$$\rho(\tilde{A}) = \lim_{\Lambda\uparrow} \rho(\tilde{A}(\Lambda)). \tag{74}$$

Here, $\tilde{A}$ should be considered as a *functional of the state*, ρ, not an element of the local algebra $\mathcal{A}$. Since $\tilde{A}$ corresponds to the global space average of A, we designate it a macroscopic observable.

According to a more general definition, due to Hepp (1972), a macroscopic observable $\tilde{A}$ is the arithmetic mean of a sequence of local observables $\{A_n\}$ for respective bounded regions $\{\Lambda_n\}$, such that (a) $\|A_n\|$ is uniformly bounded w.r.t. n, and (b) the minimum distance, d_n, of Λ_n from the origin tends to infinity with n. Thus

$$\rho(\tilde{A}) = \lim_{N\to\infty} N^{-1} \sum_{1}^{N} \rho(A_n). \tag{75}$$

One sees easily that this definition of $\tilde{A}$ reduces to the form given by (73) and (74) when $A_n = \int_{C_n} \mathrm{d}x\gamma(x)A$, where A is a bounded local observable and $\{C_n\}$ is a sequence of congruent cubes into which the space X is resolved.

The definition (75) of macroscopic observables may be employed to classify islands of states by their macroscopic properties. For, as proved by Hepp (1972), if ρ is a state with short-range correlations, in the sense of the formula (67), then any macroscopic observable $\tilde{A}$ takes the same expectation value for all states in the island $\mathscr{I}(\rho)$. This result, which we take to signify that the states in $\mathscr{I}(\rho)$ are *macroscopically equivalent*, is essentially due to the fact that the short-range correlation property of ρ ensures that its local modifications do not affect the state in arbitrarily remote spatial regions and hence have no effect on the expectation values of the macroscopic observables. A converse result, which we shall prove in Appendix B, is that if the islands, $\mathscr{I}(\rho)$ and $\mathscr{I}(\rho')$, of two states with short-range correlations, ρ and ρ', have no states in common, then there is some macroscopic observable, $\tilde{A}$, that distinguishes between them, in that $\rho(\tilde{A}) \neq \rho'(\tilde{A})$. Thus, we arrive at the conclusion that the states in the island of one with short-range correlations are macroscopically equivalent, whereas states in disjoint islands are macroscopically different from one another.

Example. The states of the model of §2.3, for which all spins are aligned, have short-range correlations; and, as effectively shown in that Section, their islands are distinguishable from one another by the fact that they yield different values for the global polarization density.

Pure phases. Since the states in the island of one with short range correlations are macroscopically equivalent, it is natural to consider such islands as candidates for *pure phases* of matter. This idea will be developed further in Chapter 4.

2.4.5. The dynamics

We shall assume that the dynamics of Σ is given, in the Schrödinger picture, by transformations $\rho \to \rho_t$ of its states such that the time-dependent expectation values of the local observables are unaffected by the matter in 'infinitely remote regions'. Thus, as the Heisenberg operator for a local observable A, evolving according to the dynamics of the finite system $\Sigma(\Lambda)$, is

$$A_{\Lambda,t} = e^{iH(\Lambda)t/\hbar} A e^{-iH(\Lambda)t/\hbar}, \tag{76}$$

we assume that

$$\rho_t(A) = \lim_{\Lambda\uparrow} \rho(A_{\Lambda,t}), \tag{77}$$

the limit being taken over an increasing sequence of bounded regions, Λ, whose union covers the space X.

The assumption of the existence of this limit certainly warrants some discussion. In fact, although the assumption has been substantiated by Streater (1967) and Robinson (1968) for lattice systems with short-range forces, e.g. the Heisenberg ferromagnet, it is clear from an argument due to Radin (1977) that the formula (77) cannot be applicable to *all* states of a continuous system. The essential point of Radin's argument is that there are certain states ρ, which he explicitly constructs, whose evolution leads to a catastrophe in which an infinite number of particles move from arbitrarily remote regions into some bounded volume, in a finite time t. When this occurs, the resultant functional ρ_t is not locally finite, and is therefore not a physical state. We shall henceforth exclude such states ρ from the model by assuming that the physical states are those that support the dynamical law given by (78), with ρ_t locally finite at all times. We shall term ρ a *stationary* state if $\rho_t = \rho$ for all t.

One can equally well describe the dynamics of Σ in the Heisenberg picture, where the evolution of the system corresponds to transformations $A \to A_t$ of the observables given formally by

$$\rho(A_t) = \rho_t(A) \equiv \lim_{\Lambda\uparrow} \rho(A_{\Lambda,t}). \tag{78}$$

Here A_t, like the macroscopic observables of the previous subsection, should be considered as a *functional of the states*, not an element of the local algebra $\mathscr{A}$. For although the local Heisenberg operator $A_{\Lambda,t}$ belongs to $\mathscr{A}$, the same is not necessarily true of A_t, as disturbances in a non-relativistic system can travel with unlimited speed and thus delocalise the time translate of a local observable.

Dynamics in an island. We shall now outline the argument establishing that, under suitable conditions, the dynamics in the island $\mathscr{I}(\rho)$ of a state ρ is of a *Hamiltonian* form; here the Hamiltonian operator acts in the GNS representation space $\mathscr{H}$ of ρ.

To formulate the dynamics in $\mathscr{I}(\rho)$, we start with the following assumptions, which are known to be satisfied by various tractable models.†

(I) If ρ' belongs to $\mathscr{I}(\rho)$, then so too does ρ'_t. Thus, the dynamics in $\mathscr{I}(\rho)$

† Cf. the models treated by Araki and Woods (1963), Streater (1967), Robinson (1968), Narnhofer (1970), and Emch and Radin (1971), among others.

corresponds to transformations $\{\bar{\alpha}_t\}$ of its states, with

$$\rho'_t = \bar{\alpha}_t \rho' \tag{79}$$

We note that, in view of the discussion of §2.4.4, this assumption, that the evolution of the system does not take its states out of $\mathcal{I}(\rho)$, signifies essentially that the dynamics in that island is sufficiently stable to ensure that local disturbances do not grow, in the course of time, to global ones.

(II) The evolution of the states in $\mathcal{I}(\rho)$ between times s and $s+t$, for arbitrary s, is also determined by the mapping $\bar{\alpha}_t$, i.e. $\rho'_{t+s} = \bar{\alpha}_t \rho'_s$. Hence, by (79), the transformations $\{\bar{\alpha}_t\}$ form a one parameter group, i.e.

$$\bar{\alpha}_t \bar{\alpha}_s = \bar{\alpha}_{t+s}. \tag{80}$$

In order to express the dynamics in $\mathcal{I}(\rho)$ in terms of a Heisenberg picture, we note that on applying the convergence condition (78) to the vector state ρ_Φ, defined by (72), it follows that $(\Phi, \hat{A}_{\Lambda,t}\Phi)$ tends to a limit as Λ increases to infinity. Since this is so for all vectors Φ in the representation space, $\mathcal{H}$, of ρ, $\hat{A}_{\Lambda,t}$ converges weakly to an operator, $\hat{A}_t$, in this limit, i.e.

$$\hat{A}_t = \text{weak} \lim_{\Lambda\uparrow} \hat{A}_{\Lambda,t}. \tag{81}$$

Hence, $\hat{A}_t$ belongs to the algebra,† $\hat{\mathcal{A}}^W$, obtained by the addition to $\hat{\mathcal{A}}$ of its weak limit points: in general, $\hat{A}_t$ does not belong to $\hat{\mathcal{A}}$ itself, since local observables of a non-relativistic system can become delocalized under time evolution.

Now it has been proved‡ by the author (Sewell 1982) that, just as the time-translations of a finite system can be equivalently expressed in terms of the Schrödinger transformations of its states or the Heisenberg automorphisms of its observables, the assumptions (I) and (II) imply the corresponding thing for the dynamics in $\mathcal{I}(\rho)$, with $\hat{\mathcal{A}}^W$ playing the role of the algebra of observables. Specifically, we have proved that, under those assumptions, the transformations $\hat{A} \to \hat{A}_t$ correspond to a one-parameter group $\{\hat{\alpha}_t \mid \hat{\alpha}_t \hat{\alpha}_s = \hat{\alpha}_{t+s}\}$ of *automorphisms of* $\hat{\mathcal{A}}^W$, with

$$\hat{\alpha}_t \hat{A} = \hat{A}_t. \tag{82}$$

Moreover, Kadison (1965) has proved that, under rather general conditions, such automorphisms are implemented by a one-parameter group $\{\hat{U}_t \mid \hat{U}_t \hat{U}_s = \hat{U}_{t+s}\}$ of transformations of $\mathcal{H}$, i.e.

$$\hat{\alpha}_t \hat{A} = \hat{U}_t \hat{A} \hat{U}_{-t}. \tag{83}$$

† A weakly closed *-algebra of bounded operators in a Hilbert space, such as $\hat{\mathcal{A}}^W$, is generally termed a *Von Neumann* algebra, or a W^*-algebra.

‡ This result generalizes an earlier one, due to Dubin and Sewell (1970), who proved the same thing for the dynamics in islands of thermal equilibrium states.

In particular, this is the case when ρ is *stationary*. For then it follows from a theorem† of Segal (1951) that the formula (83) is valid, with the unitary transformations defined by the equation

$$\hat{U}_t\hat{A}\Psi = (\hat{\alpha}_t\hat{A})\Psi \equiv \hat{A}_t\Psi \quad \text{for all } A \text{ in } \mathscr{A}, \tag{84}$$

Ψ being the cyclic vector for ρ, as specified in §2.4.3. By (61), (70), (80), and (84), $\hat{U}_t$ is equivalently given by the formula

$$(\hat{B}\Psi, \hat{U}_t\hat{A}\Psi) = \lim_{\Lambda\uparrow} \mathrm{Tr}_\Lambda(\rho_\Lambda A_{\Lambda,t}). \tag{85}$$

An argument,‡ due to Winnink (1973), shows that this formula, together with the local finiteness of ρ, implies that $\hat{U}_t$ is continuous in t – as for a finite system. Therefore, by Stone's theorem,§ the one-parameter group $\{\hat{U}_t\}$ has an infinitessimal generator $\mathrm{i}\hat{H}/\hbar$, with $\hat{H}$ self-adjoint, i.e.

$$\hat{U}_t = \mathrm{e}^{\mathrm{i}\hat{H}t/\hbar}. \tag{86}$$

We term $\hat{H}$ the Hamiltonian for the island $\mathscr{I}(\rho)$. One sees immediately from (82), (83), and (86) that the operators $\hat{A}_t$ satisfy the *Heisenberg equation* of motion corresponding to this Hamiltonian, i.e.

$$\frac{\mathrm{d}\hat{A}_t}{\mathrm{d}t} = \frac{\mathrm{i}}{h}[\hat{H}, \hat{A}_t]. \tag{87}$$

Thus, the dynamics in the island of a stationary state, ρ, is governed by the Hamiltonian $\hat{H}$ in the representation space $\mathscr{H}$. This operator is evidently a *representation-dependent* quantity, its form being governed by the state ρ as well as the interactions in the system. A good illustration of this is provided by the example of an Ising–Weiss ferromagnetic model. For Emch and Knops (1970) have shown that the Hamiltonian for the island of an equilibrium state of this model contains a contribution due to the interactions of the spins with the temperature-dependent 'inner field', arising from the global polarization of the system.

Absence of Poincaré cycles. We saw in §2.2.4 that the dynamics of a finite system is quasi-periodic, due to the discreteness of the spectrum of its Hamiltonian. For the same reason, the time-dependent local observables $A_{\Lambda,t}$, defined by (76), are quasi-periodic. However, as the discreteness of

† The proof is essentially that (a) defining $\{\hat{U}_t\}$ by (84), this set of operators inherits the one-parameter group property from $\{\hat{\alpha}_t\}$; (b) for stationary ρ, $(\hat{A}_t\Psi, \hat{B}_t\Psi) \equiv (\hat{A}\Psi, \hat{B}\Psi)$ and thus, by (84), $\hat{U}_t{}^*\hat{U}_t = I$, which implies the unitarity of the group $\{U_t\}$; and (c) by (85) and the automorphic property of $\hat{\alpha}_t$, $\hat{U}_t\hat{A}\hat{U}_{-t}\hat{B}\Psi = \hat{U}_t\hat{A}\hat{B}_{-t}\Psi = (\hat{A}\hat{B}_{-t})_t\Psi = \hat{A}_t\hat{B}\Psi$ and therefore (83) is valid.

‡ The argument is that the LHS of (85) is the limit of a sequence of continuous functions of t, from which it follows that $\hat{U}_t$ is measurable w.r.t. t. Since the GNS space $\mathscr{H}$ of a locally finite state ρ is known to be separable (cf. Hugenholtz and Wieringa 1969), this in turn implies that $\hat{U}_t$ is continuous in t (cf. Riesz and Sz-nagy 1955, p. 387).

§ Cf. Appendix A.

the spectrum of $H(\Lambda)$ generally arises from finite-size effects connected with the conditions at the boundary of Λ, one anticipates that this spectrum becomes continuous in the limit where Λ becomes infinite, and thus that the time-dependent expectation values, $\rho_t(A)$, of the observables of Σ, as given by (77), are not quasi-periodic. In fact, this has been proved to be the case for various tractable models,† and one envisages that it is generically true for infinite systems. This is of crucial importance for the theory of non-equilibrium phenomena. For assuming Σ to be free from Poincaré cycles, at least in suitable islands of states, the model is amenable to a systematic theory of irreversible processes, in which dynamical variables relax into some terminal values as the time, t, tends to infinity. Examples of such processes will be given in Chapter 7.

2.4.6. *Unbounded local observables*

The algebraic structure of the bounded local observables has been exploited in §§2.4.3–4 for the purposes of formulating their representations and of classifying the associated islands of states according to their macroscopic properties. Once this has been achieved, there is no difficulty in extending the model to the unbounded local observables.‡ We shall now sketch how this has been done (Sewell 1970).

We take the unbounded observables, A, for the bounded region Λ to be self-adjoint operators in $\mathcal{H}(\Lambda)$ such that the bounded functions of A belong to $\mathcal{A}(\Lambda)$: thus, if $\mathcal{A}(\Lambda)$ is formulated in second quantization, as in Example 2 of §2.4.1, then the unbounded local observables for the region Λ are the self-adjoint, gauge invariant operators in $\mathcal{H}(\Lambda)$. As in our treatment of the unbounded observables of a finite system in §2.2, we may resolve any unbounded observable, A, for the region Λ, into parts A_N and A'_N, whose spectra lie inside and outside the interval $[-N, N]$, respectively, and then express A as the limit of A_N (cf. (21)), i.e.

$$A = \lim_{N\to\infty} A_N. \tag{88}$$

The expectation value, $\rho(A)$, of A for the state ρ is defined to be $\mathrm{Tr}_\Lambda(\rho_\Lambda A)$, where, as previously, ρ_Λ is the density matrix representing ρ in the region Λ. Thus, by (61), this expectation value is equal to $\lim_{N\to\infty} \mathrm{Tr}_\Lambda(\rho_\Lambda A_N) \equiv \lim_{N\to\infty} \rho(A_N)$, which means that the expectation value of the unbounded observable A is the limit of that of the bounded one, A_N, as $N \to \infty$, i.e.

$$\rho(A) = \lim_{N\to\infty} \rho(A_N). \tag{89}$$

† See, for example, the models treated by Araki and Woods (1963), Narnhofer (1970), Emch and Radin (1971), Hepp (1972), and Sewell (1974).

‡ Alternatively, one may formulate the model, *ab initio*, in terms of a suitably constructed algebra of unbounded observables (cf. Borchers 1962, 1973; Uhlmann 1962; Powers 1971, 1974; and Alcantara and Dubin 1981).

Similarly, the GNS representation for the state ρ may be extended in a consistent† way, to the unbounded observables, A, by the formula

$$\hat{A} = \lim_{N \to \infty} \hat{A}_N, \tag{90}$$

where the domain of $\hat{A}$ consists of the vectors in $\mathscr{H}$ on which $\hat{A}_N$ converges strongly as $N \to \infty$. Correspondingly, the Heisenberg equation of motion (87) may be extended to the representation of the unbounded observables for the island of a stationary state. Thus, the main features of the model are generalized to the unbounded observables.

2.4.7. *Summary of the model*

The mathematical formulation of the infinite system Σ may be summarized as follows. The *bounded local observables* are the self-adjoint elements of the algebra $\mathscr{A}$, given by the union of the local algebras, $\mathscr{A}(\Lambda)$. The *states* are the locally finite, linear, positive, normalized functionals on $\mathscr{A}$, as defined by eqn (61). These may be classified according to their macroscopic properties, since the states in an island corresponding to a pure phase as globally equivalent to one another. The *dynamics* is given by a limiting form of that of a corresponding finite system (cf. eqn (78)). In particular, this reduces to a *Hamiltonian* form in the island of a stationary state, though now the Hamiltonian itself depends not only on the forces but also on the macroscopic properties of the island. Finally, the model is extended by the formulae (88)–(90) to the *unbounded local observables*.

2.5. Concluding remarks

We have seen in the preceding Sections that the idealization, whereby a macroscopic system is represented as infinite, provides new structures, which form a natural framework for theories of collective phenomena. Among the noteworthy features of these structures are the following.

(1) The macroscopic and microscopic descriptions of the model are quite distinct from one another. In particular, the macroscopic specification of a state serves to determine the island in which it lies, while the microscopic one identifies it as a definite element of that island (cf. §§2.3, 2.4.4).

(2) The model admits a clear-cut characterization of symmetry breakdown in phase transitions, as illustrated by the simple example of §2.3. More generally, it leads to a description of a phase as a class of macroscopically equivalent states (i.e. an island) associated with the GNS representation of one with short-range correlations (cf. §2.4.4). This description will be developed in a thermodynamical context in Chapters 3 and 4.

† It may be seen from our constructions in the above-cited article that this consistency depends on the local finiteness of ρ.

(3) The dynamics of an infinite system, unlike that of a finite one, is generally free from Poincaré cycles, and thus the model is amenable to a systematic theory of irreversible processes. Furthermore, the dynamics in different islands may, in general, be quite different from one another, in accordance with the empirical fact that the non-equilibrium properties of a macroscopic system generally depend on which phase it is in (cf. §2.4.5).

Appendix A
Hilbert spaces

Hilbert spaces are generalizations of Euclidean spaces to arbitrary dimensionality, which may be infinite. Authoritative treatises on these spaces are given in the book by Riesz and Nagy (1955) on Functional Analysis and in Von Neumann's (1955) book on Quantum Mechanics. Here we shall present a rudimentary account of the theory of these spaces, which should suffice for the purposes of this book.

A *Hilbert space* is a set, $\mathcal{H}$, of elements, or vectors, $(f, g, h, \ldots)$ which satisfies the following conditions (1)–(5).

(1) If f and g belong to $\mathcal{H}$, then there is a unique element of $\mathcal{H}$, denoted by $f+g$, the operation of addition $(+)$ being invertible, commutative and associative.

(2) If λ is a complex number, then for any f in $\mathcal{H}$, there is an element λf of $\mathcal{H}$; and the multiplication of vectors by complex numbers thereby defined satisfies the distributivity conditions

$$\lambda(f+g) = \lambda f + \lambda g; \quad \text{and} \quad (\lambda + \mu)f = \lambda f + \mu f.$$

(3) $\mathcal{H}$ possess a zero element, 0, characterized by the property that

$$0 + f = f$$

for all vector f in $\mathcal{H}$.

(4) For each pair of vectors f, g in $\mathcal{H}$, there is a complex number (f, g), termed the *inner product* or scalar product, of f with g, such that

$$(f, g) = \overline{(g, f)}$$

$$(f, g+h) = (f, g) + (f, h)$$

$$(f, \lambda g) = \lambda(f, g)$$

and

$$(f, f) \geqslant 0,$$

equality in the last formula occurring only if $f=0$. We define $\|f\|$, the *norm* of f, as the non-negative square root of (f,f), i.e.

$$\|f\| = (f,f)^{\frac{1}{2}}. \tag{A.1}$$

It follows from these specifications that the inner product and norm satisfy the Schwartz inequality

$$|(f,g)| \leqslant \|f\|\,\|g\| \tag{A.2}$$

and the triangular inequality

$$\|f-g\| \leqslant \|f\| + \|g\|. \tag{A.3}$$

In view of the latter inequality, one may regard $\|f-g\|$ as the 'distance' between f and g.

(5) If $\{f_n\}$ is a sequence in $\mathscr{H}$ satisfying the Cauchy condition that $\|f_m - f_n\| \to 0$ as m, n tend independently to infinity, then there is a unique element f of $\mathscr{H}$ such that $\|f_n - f\| \to 0$ as $n \to \infty$. $\mathscr{H}$ is termed *complete* by virtue of this property.

The following are examples of Hilbert spaces.

***Example* 1.** The N-dimensional space consisting of elements f with complex coordinates $(f_1, \ldots, f_N)$ and inner product defined by

$$(f,g) = \sum_1^N \bar{f}_n g_n.$$

***Example* 2** is the infinite-dimensional counterpart of the previous example. Thus, its elements, f, are sequences of complex numbers $(f_1, \ldots, f_n, \ldots)$ such that $\Sigma_1^\infty |f_n|^2$ is finite, and its inner product is given by

$$(f,g) = \sum_1^\infty \bar{f}_n g_n.$$

***Example* 3** consists of the set of complex functions, f, of a real variable x such that $\int_{-\infty}^{\infty} \mathrm{d}x\, |f(x)|^2$ is finite, with inner product given by

$$(f,g) = \int_{-\infty}^{\infty} \mathrm{d}x \bar{f}(x) g(x).$$

Strong and weak convergence. A sequence $\{f_n\}$ of vectors in a Hilbert space $\mathscr{H}$ is said to converge *strongly* to f if $\|f_n - f\| \to 0$ as $n \to \infty$. On the other hand, the sequence is said to converge *weakly* to f if $(f_n, g) \to (f, g)$ as $n \to \infty$, for any vector g in $\mathscr{H}$. It follows immediately from these definitions and the Schwartz inequality (A.2) that strong convergence implies weak convergence. The converse is not generally true, however.

Dense sets of vectors. A set, Δ, of vectors in a Hilbert space $\mathscr{H}$ is said to be dense in $\mathscr{H}$ if any vector f in $\mathscr{H}$ is the strong limit of some sequence $\{f_n\}$ in

Δ. This condition may be equivalently stated by saying that, for each vector f in $\mathcal{H}$ and each positive number ε, there is a vector f' in Δ such that $\|f'-f\|<\varepsilon$.

Orthonormality. A vector, f, in $\mathcal{H}$, is referred to as *normalized* if $\|f\|=1$. Two vectors f and g, in $\mathcal{H}$, are termed *orthogonal* if $(f,g)=0$. A family of vectors $\{\phi_\alpha\}$, with α running through some index set, is termed *orthonormal* if its members are normalized and mutually orthogonal.

Separability. A Hilbert space $\mathcal{H}$ is termed *separable* if it contains an orthonormal *sequence*† $\{\phi_n\}$ such that any vector f in $\mathcal{H}$ may be expressed as a linear combination of the ϕ_n's, i.e.

$$f=\sum_1^\infty c_n\phi_n \equiv \text{strong}\lim_{N\to\infty}\sum_1^N c_n\phi_n,$$

where the c_n's are complex numbers. In this case, the sequence $\{\phi_n\}$ is termed an *orthonormal basis* in $\mathcal{H}$.

We shall henceforth assume that $\mathcal{H}$ is separable, since this is the case for the Hilbert spaces we are concerned with in this book.

Subspaces. A subspace $\mathcal{K}$ of $\mathcal{H}$ is a subset‡ of vectors in which themselves form a Hilbert space, in that they satisfy the above conditions (1)–(5). It follows from this definition that, if $\mathcal{K}$ is a subspace of $\mathcal{H}$, then so too is the set, $\mathcal{K}^\perp$, of vectors orthogonal to all those in $\mathcal{K}$. The subspace $\mathcal{K}^\perp$ is termed the *orthogonal complement* of $\mathcal{K}$ in $\mathcal{H}$. Further, any vector, f, in $\mathcal{H}$ may be uniquely decomposed into components $f_\mathcal{K}$ and $f_{\mathcal{K}^\perp}$, lying in $\mathcal{K}$ and $\mathcal{K}^\perp$, respectively, i.e.

$$f=f_\mathcal{K}+f_{\mathcal{K}^\perp}. \tag{A.4}$$

Here, $f_\mathcal{K}$ is the vector f' in $\mathcal{K}$ for which $\|f-f'\|$ takes its minimum value.

Example. If $\mathcal{H}$ is the Hilbert space specified in Example 2, above, then the vectors $f=(f_1,\ldots,f_N,0,\ldots,0\ldots)$, with $f_n=0$ for $n>N$, form a subspace $\mathcal{K}$. The orthogonal complement, $\mathcal{K}^\perp$, of $\mathcal{K}$ then consists of the vectors $(0,\ldots,0,f_{N+1},\ldots)$, with $f_n=0$ for $n\leq N$.

Linear manifolds. A linear manifold, $\mathcal{M}$, of a Hilbert space, $\mathcal{H}$, is a set of vectors in $\mathcal{H}$ such that if f and g belong to $\mathcal{M}$, then so too do $(f+g)$ and λf, for any complex number λ. Thus, a linear manifold, $\mathcal{M}$, is a subspace of $\mathcal{H}$ *if* it contains the limits of its Cauchy sequences (cf. condition (5) for the

† In the case of a non-separable Hilbert space, one needs instead an orthonormal basis $\{\phi_\alpha\}$ with α running through a *non-denumerable* index set.

‡ We take it that all the subsets, or subspaces, $\mathcal{K}$ that we deal with are *proper*, i.e. that they neither reduce to the zero vector only, nor constitute the whole of $\mathcal{H}$.

definition of a Hilbert space). In any case, the manifold, $\bar{\mathcal{M}}$, obtained by adding these limit points to $\mathcal{M}$, is a subspace of $\mathcal{H}$. $\bar{\mathcal{M}}$ is called the closure of $\mathcal{M}$.

Operators in a Hilbert space. An operator, A, in a Hilbert space, $\mathcal{H}$, is a linear transformation of a linear manifold, $\mathcal{D}(A)$ ($\subset \mathcal{H}$), into $\mathcal{H}$. The manifold $\mathcal{D}(A)$ is termed the domain of definition, or simply the *domain*, of A.

The operator A is termed *bounded* if the set of numbers, $\|Af\|$, is bounded as f runs through the normalized vectors in $\mathcal{D}(A)$. In this case, we define $\|A\|$, the *norm* of A, to be the supremum, i.e. the least upper bound, of $\|Af\|$, as f runs through these normalized vectors. It follows from this definition that

$$\|Af\| \leqslant \|A\| \, \|f\| \text{ for all vectors } f \text{ in } \mathcal{D}(A). \tag{A.5}$$

Note on domains. If A is bounded, we may take $\mathcal{D}(A)$ to be $\mathcal{H}$, since, even if this domain is originally defined to be a proper subset of $\mathcal{H}$, we can always extend it to the whole of this space by the following procedure. First, noting that $\|Af_m - Af_n\| \leqslant \|A\| \, \|f_m - f_n\|$, we infer that the convergence of a sequence of vectors $\{f_n\}$ in $\mathcal{D}(A)$ implies that of $\{Af_n\}$. Hence, we may extend the definition of A to $\overline{\mathcal{D}(A)}$, the closure of $\mathcal{D}(A)$, by defining $A \lim_{n\to\infty} f_n$ to be $\lim_{n\to\infty} Af_n$. We may then extend A to the full Hilbert space $\mathcal{H}$, by defining it to be zero on $\overline{\mathcal{D}(A)}^{\perp}$, the orthogonal complement of $\overline{\mathcal{D}(A)}$.

On the other hand, if A is unbounded, then, in general, $\mathcal{D}(A)$ does not comprise the whole of $\mathcal{H}$ and cannot be extended to do so. This is exemplified by the differential operator $\mathrm{d}/\mathrm{d}x$ acting on the space, $\mathcal{H}$, of square-integrable functions of the real variable x. For the domain of this operator consists of those functions $f(x)$ for which both $\int \mathrm{d}x \, |f(x)|^2$ and $\int \mathrm{d}x \, | \mathrm{d}f(x)/\mathrm{d}x|^2$ are both finite, and this set of functions does not comprise the whole of $\mathcal{H}$.

Adjoint of an operator. In the case where A is a bounded operator in $\mathcal{H}$, we define A^*, the *adjoint* of A, by the formula

$$(f, A^*g) = (Af, g) \text{ for all } f, g \text{ in } \mathcal{H}. \tag{A.6}$$

The operator A is then termed *self-adjoint* if $A^* = A$ or, equivalently, if

$$(f, Ag) = (Af, g) \text{ for all } f, g \text{ in } \mathcal{H}. \tag{A.7}$$

In the case where A is an unbounded operator in $\mathcal{H}$ we again define its adjoint, A^*, by the same formula, except that f is confined to $\mathcal{D}(A)$ and g to the domain $\mathcal{D}(A^*)$, which is specified as follows: g belongs to $\mathcal{D}(A^*)$ if there is a vector g_A in $\mathcal{H}$ such that $(f, g_A) = (Af, g)$ for all f in $\mathcal{D}(A)$, in which case $g_A = A^*g$. The operator A is termed self-adjoint if $\mathscr{D}(A^*) = \mathscr{D}(A)$ and $A^* = A$: the coincidence of $\mathscr{D}(A^*)$ with $\mathscr{D}(A)$ is essential here. The domain of a self-adjoint operator is dense in $\mathscr{H}$.

Positive operators. A self-adjoint operator A is termed positive if $(f, Af) \geqslant 0$ for all vectors f in $\mathscr{D}(A)$. Thus, for example, if $A = B^*B$ where B is a bounded operator, then it follows from the above definitions that A is positive.

Algebraic properties of the operator norm. It follows from the definitions of the norm and the adjoint of a bounded operator, together with the triangular inequality (A.3) that if A, B are bounded operators and λ is a complex number, then

$$\left.\begin{array}{l} \|\lambda A\| = |\lambda|\,\|A\|; \qquad \|A^*A\| = \|A\|^2; \qquad \|A+B\| \leqslant \|A\| + \|B\|; \\ \text{and} \quad \|AB\| \leqslant \|A\|\,\|B\|. \end{array}\right\} \tag{A.8}$$

Projection operators. Suppose that $\mathscr{K}$ is a subspace of $\mathscr{H}$. Then since any vector f in $\mathscr{H}$ may be resolved into unequally defined components $f_{\mathscr{K}}$ and $f_{\mathscr{K}^\perp}$ in $\mathscr{K}$ and $\mathscr{K}^\perp$, respectively, according to (A.4), we may define an operator P by the formula $Pf = f_{\mathscr{K}}$. This is termed the *projection operator* from $\mathscr{H}$ to $\mathscr{K}$, or simply the projection operator or projector for the subspace $\mathscr{K}$. It follows from this definition and the orthogonality of $f_{\mathscr{K}}$ and $f_{\mathscr{K}^\perp}$ that $\|Pf\|^2 = \|f_{\mathscr{K}}\|^2 \leqslant \|f_{\mathscr{K}}\|^2 + \|f_{\mathscr{K}^\perp}\|^2 = \|f\|^2$, and therefore that P is bounded. It also follows from the definition of P that

$$P^2 = P = P^*. \tag{A.9}$$

In fact, this formula is generally employed as a *definition* of a projection operator, P, since it implies that the set of elements $\{Pf\}$ form a subspace of $\mathscr{H}$, as f runs through the vectors in $\mathscr{H}$.

We note that, in the case where $\mathscr{K}$ is a one-dimensional subspace, consisting of the scalar multiples of a normalized vector ϕ, the projection operator $P = P(\phi)$ is given by

$$P(\phi)f = (\phi, f)\phi. \tag{A.10}$$

Unitary operators. An operator, U, in a Hilbert space $\mathscr{H}$ is termed unitary if $(Uf, Ug) = (f, g)$ for all vectors f, g in $\mathscr{H}$, and if U has an inverse U^{-1}, i.e. $UU^{-1} = U^{-1}U = I$, the identity operator ($If \equiv f$). In other words, a unitary operator is an invertible one, which preserves the form of the inner product in $\mathscr{H}$. The above definition of unitarity is equivalent to the condition that

$$U^*U = UU^* = I; \quad \text{i.e.} \quad U^* = U^{-1}. \tag{A.11}$$

A unitary mapping of $\mathscr{H}$ onto a second Hilbert space $\mathscr{H}'$ is an invertible transformation V, from $\mathscr{H}$ to $\mathscr{H}'$, such that $(f, g)_{\mathscr{H}} \equiv (Vf, Vg)_{\mathscr{H}'}$.

Operator convergence. Suppose that A and the sequence $\{A_n\}$ are bounded operators in $\mathscr{H}$. Then A_n is said to converge *uniformly*, or *in norm*, to A as $n \to \infty$ if $\|A_n - A\| \to 0$ as $n \to \infty$; A_N is said to converge *strongly* to A if $A_n f$ tends strongly to Af for all vectors f in $\mathscr{H}$; and A_n is said to converge *weakly* to A if $A_n f$ tends weakly to Af for all vectors f in $\mathscr{H}$. From these definitions, it follows that norm convergence implies strong operator convergence, which in turn implies weak operator convergence. The converse statements are applicable only if $\mathscr{H}$ is finite dimensional.

Density matrices and Traces. A density matrix, ρ, is an operator in $\mathscr{H}$ of the form

$$\rho = \sum_{1}^{\infty} w_n P(\phi_n) = \text{norm} \lim_{N \to \infty} \sum_{1}^{N} w_n P(\phi_n), \tag{A.12}$$

where $\{P(\phi_n)\}$ are the projection operators for an orthonormal sequence $\{\phi_n\}$ of vectors and $\{w_n\}$ is a sequence of non-negative numbers whose sum is unity. Thus, a density matrix is bounded and positive.

The *Trace* of a positive operator B is defined to be

$$\mathrm{Tr}(B) = \sum_{n} (\phi_n, B\phi_n), \tag{A.13}$$

where $\{\phi_n\}$ is any orthonormal basis set; the value of $\mathrm{Tr}(B)$, which is infinite for some operators, is independent of the choice of basis.

It follows from these definitions of density matrices and Trace that a density matrix is a positive operator whose Trace is equal to unity.

One-parameter unitary groups: Stone's theorem. A one-parameter group of unitary transformations of $\mathscr{H}$ is a family $\{U_t\}$ of unitary operators in $\mathscr{H}$, with t running through the real numbers, such that

$$U_t U_s = U_{t+s}; \qquad U_0 = I. \tag{A.14}$$

The group is said to be *continuous* if U_t converges strongly to I as t tends to zero; or equivalently, if U_t converges strongly to U_{t_0} as t tends to t_0, for any real t_0. In this case, Stone's theorem tells us that there is a unique self-adjoint operator, K, in $\mathscr{H}$ such that

$$\frac{\mathrm{d}}{\mathrm{d}t} U_t f = \mathrm{i} K U_t f = \mathrm{i} U_t K f \quad \text{for all } f \text{ in } \mathscr{D}(K). \tag{A.15}$$

This equation is formally expressed as

$$U_t = \mathrm{e}^{\mathrm{i}Kt}, \tag{A.16}$$

and $\mathrm{i}K$ is termed the *infinitesimal generator* of the group $\{U_t\}$.

Spectral analysis. Suppose that A is a self-adjoint, possibly unbounded, operator in $\mathscr{H}$ and ϕ is a vector in $\mathscr{H}$, such that $A\phi = \lambda\phi$, where λ is a number, then ϕ is termed an eigenvector and λ the corresponding eigenvalue of A: the self-adjointness of A ensures that λ is real. The operator A is said to have a *discrete spectrum* if it has a set of eigenvectors $\{\phi_n\}$ which form a complete orthonormal basis in $\mathscr{H}$. In this case, A may be expressed in the form

$$A = \sum \lambda_n P(\phi_n), \tag{A.17}$$

where $P(\phi_n)$ is the projection operator and λ_n the eigenvalue for ϕ_n.

In general, even when A does not have a discrete spectrum, it may still be resolved into a linear combination of projection operators according to the *spectral theorem* which serves to express A as a Stieltjes integral,

$$A = \int \lambda \, \mathrm{d}E(\lambda), \tag{A.18}$$

where $\{E(\lambda)\}$ is a family of intercommuting projectors, such that $E(-\infty) = 0$, $E(\infty) = I$, $E(\lambda) \leqslant E(\lambda')$ if $\lambda < \lambda'$, and $E(\lambda')$ converges strongly to $E(\lambda)$ as λ' tends to λ from above. Here $E(\lambda)$ is a function of A, i.e. $\chi_\lambda(A)$, where $\chi_\lambda(x)$ is the function of the real variable, x, given by

$$\chi_\lambda(x) = \left.\begin{matrix} 1 & \text{for } x < \lambda. \\ 0 & \text{for } x \geqslant \lambda. \end{matrix}\right\}$$

In the particular case where A has a discrete spectrum, i.e. $A = \sum \lambda_n P(\phi_n)$, then $E(\lambda) = \sum_{\lambda_n < \lambda} P(\phi_n)$.

In general, it follows from the spectral theorem that, for any positive N, we may express A in the form

$$A = A_N + A'_N \tag{A.19}$$

where

$$A_N = \int_{-N-0}^{N+0} \lambda \, \mathrm{d}E(\lambda) \quad \text{and} \quad A'_N = \int_{-\infty}^{-N-0} \lambda \, \mathrm{d}E(\lambda) + \int_{N+0}^{\infty} \lambda \, \mathrm{d}E(\lambda). \tag{A.20}$$

Thus, A is decomposed into parts, A_N and A'_N, whose spectra lie inside and outside the interval $[-N, N]$, respectively, and

$$A = \lim_{N \to \infty} A_N \tag{A.21}$$

on the domain of A. The significance of this last formula is that it expresses unbounded operators as limits of bounded ones.

Tensor products. Suppose that $\mathscr{H}_1$ and $\mathscr{H}_2$ are Hilbert spaces, and that $\mathscr{H}$ is a third Hilbert space, defined in terms of $\mathscr{H}_1$ and $\mathscr{H}_2$ by the following specifications.

(a) For each pair of vectors f_1, f_2 in $\mathscr{H}_1$, $\mathscr{H}_2$, respectively, there is a vector in

$\mathscr{H}$, denoted by $f_1 \otimes f_2$, such that

$$(f_1 \otimes f_2, g_1 \otimes g_2) = (f_1, g_1)_{\mathscr{H}_1}(f_2, g_2)_{\mathscr{H}_2}.$$

(b) $\mathscr{H}$ consists of the linear combinations of the vectors $f_1 \otimes f_2$, together with the strong limits of their Cauchy sequences. We term $\mathscr{H}$ the *tensor product* of $\mathscr{H}_1$ and $\mathscr{H}_2$, and denote it by $\mathscr{H}_1 \otimes \mathscr{H}_2$.

If A_1 and A_2 are operators in $\mathscr{H}_1$ and $\mathscr{H}_2$, respectively, we define the operator $A_1 \otimes A_2$ in $\mathscr{H}_1 \otimes \mathscr{H}_2$ by the formula

$$(A_1 \otimes A_2)(f_1 \otimes f_2) = (A_1 f_1) \otimes (A_2 f_2).$$

$A_1 \otimes A_2$ is called the *tensor product* of A_1 and A_2.

Similarly, we may define the tensor product $\mathscr{H}_1 \otimes \mathscr{H}_2 \ldots \otimes \mathscr{H}_n$ as well as that, $A_1 \otimes A_2 \ldots \otimes A_n$, of operators $A_1, \ldots, A_n$. In a standard notation, one writes

$$\bigotimes_{j=1}^{n} \mathscr{H}_j = \mathscr{H}_1 \otimes \mathscr{H}_2 \ldots \otimes \mathscr{H}_n; \quad \text{and} \quad \bigotimes_{j=1}^{n} A_j = A_1 \otimes A_2 \ldots \otimes A_n.$$

Appendix B
Macroscopic observables and islands of states

We shall now prove a result stated in §2.4.4, namely that if ρ and ρ' are states with short-range correlations, and if the islands $\mathscr{I}(\rho)$ and $\mathscr{I}(\rho')$ are disjoint from one another, then there is some macroscopic observable $\tilde{A}$ such that $\rho(\tilde{A}) \neq \rho'(\tilde{A})$.

The proof is based on a simple adaptation of a theorem of Ruelle (1969b) which established that, under the given assumptions on ρ and ρ', any mixture $\lambda\rho + (1-\lambda)\rho'$ of these states does *not* have short range correlations. Thus, there exists a local observable B and a sequence $\{A_n\}$, conforming to the specifications of §2.4.4, such that, if

$$\tilde{\rho} = \tfrac{1}{2}(\rho + \rho'), \tag{B.1}$$

then

$$\tilde{\rho}(BA_n) - \tilde{\rho}(B)\tilde{\rho}(A_n) \text{ does not tend to zero as } n \to \infty. \tag{B.2}$$

Furthermore, as $\{\|A_n\|\}$ is uniformly bounded, we may assume that† $\rho(A_n)$ and $\rho'(A_n)$ converge to limits a and a', say, as $n \to \infty$, since this can always be ensured by passing to a subsequence. Hence, it follows from the definition of $\tilde{A}$ (i.e. eqn (76)) that

$$\rho(\tilde{A}) = a \quad \text{and} \quad \rho'(\tilde{A}) = a', \tag{B.3}$$

while it follows from the fact that ρ and ρ' have short-range correlations

† This is justified by the Bolzano–Weierstrass Theorem (cf. Whittaker and Watson 1946, p. 12).

that

$$\rho(BA_n) \to a\rho(B) \quad \text{and} \quad \rho'(BA_n) \to a'\rho(B) \text{ as } n \to \infty. \tag{B.4}$$

Hence, by eqns (B.1, 3, 4),

$$\tilde{\rho}(BA_n) - \tilde{\rho}(B)\tilde{\rho}(A_n) \to \tfrac{1}{4}(\rho(\tilde{A}) - \rho'(\tilde{A}))(\rho(B) - \rho'(B)) \text{ as } n \to \infty. \tag{B.5}$$

On comparing this formula with (B.2), one sees immediately that $\rho(\tilde{A}) \neq \rho'(\tilde{A})$, as required.

Part II

Statistical thermodynamics

3

Equilibrium states

3.1. Introduction

The traditional form of Statistical Thermodynamics† is based on the standard Gibbs ensembles of large, but finite, systems. Although that theory has been enormously successful in relating the equilibrium bulk properties of matter to the laws of microphysics, it does suffer from certain basic deficiencies. For example, it cannot accommodate different phases of a system (e.g. liquid and vapour) under the same thermodynamical conditions, since the Gibbs ensemble representing the equilibrium state of a *finite* system is uniquely determined by the prevailing macroscopic constraints: thus, if the volume, temperature and mass are controlled to take specified values, then the resultant ensemble is the canonical one. Furthermore, as emphasized by Fisher (1964), the traditional theory cannot provide a description of metastable states, such as superheated or supercooled phases of matter. The essential reason is that, if ΔF is the difference between the free energies of metastable and equilibrium states of a system at temperature T, then the probability of occurrence of the former state in a Gibbs ensemble is $\exp(-\Delta F/kT)$; and, in view of the extensivity of ΔF, this probability is negligibly small, decreasing exponentially with the size of the system.

It is therefore evident that the standard statistical mechanics of finite systems does not possess sufficient structure to provide a general theory either of thermodynamic phases or of the closely related phenomenon of metastable states. On the other hand, we have an indication that a statistical mechanical theory, based on *infinite* systems, might possess the required structure, since, as observed in the previous chapter (§2.3), an infinite Heisenberg ferromagnetic chain supports a number of different phases, distinguished by the directions of their polarizations, at absolute zero. Our objective now is to formulate the statistical mechanics of infinite systems in a way that leads to a general theory of thermodynamic phases. We must first emphasize, however, that this cannot be achieved by a trivial extrapolation of the standard Gibbs ensemble theory for finite systems, for the following reasons.

(a) The canonical density matrix of a *finite* system takes the form

$$\rho_c = \exp - \beta(H - \Phi(T)), \tag{1}$$

† For an authoritative account of traditional Statistical Thermodynamics, see, for example, the books of Schrödinger (1952) and Landau and Lifshitz (1959).

where

$$\beta = (kT)^{-1}, \tag{2}$$

k is Boltzmann's constant, H the Hamiltonian and

$$\Phi(T) = -kT \ln \mathrm{Tr}\, \mathrm{e}^{-\beta H}, \tag{3}$$

the free energy. The formula (1) could not be extrapolated to an infinite system, since the extensivity of the free energy would then render $\Phi(T)$ infinite.

(b) As we saw in Chapter 2, the Hilbert space representation of the observables of an infinite system is not given a priori, but depends on the prevailing macroscopic conditions.

In view of these considerations, we seek a generalization of statistical thermodynamics to infinite systems, that is based on thermodynamical and dynamical stability properties of canonical states, rather than on a direct extrapolation of the formula (1). Thus, in order to obtain such a generalization, we first note that the canonical state, ρ_c, of a finite system is uniquely determined by the *thermodynamical stability condition* that it minimizes the free energy functional

$$\hat{F}(\rho) = \mathrm{Tr}(\rho H + kT\,\rho \ln \rho): \tag{4}$$

the proof that this condition is equivalent to the formula (1) will be given in Appendix A. In view of eqns (1)–(4), this thermodynamical stability condition may be expressed in the following form:

$$\hat{F}(\rho)\begin{cases} = \Phi(T) & \text{for} \quad \rho = \rho_c \\ > \Phi(T) & \text{for} \quad \rho \neq \rho_c \end{cases}. \tag{5}$$

Another way of characterizing ρ_c stems from the *fluctuation–dissipation theorem*,† due to Callen and Welton (1951), which relates the equilibrium fluctuations of a system to its linear response to applied forces. An elegant form of this theorem, due to Kubo (1957), may be described as follows. Suppose that a finite system, Σ, is in its canonical state, ρ_c, at time $t = 0$, and is subsequently subjected to a Hamiltonian perturbation $BF(t)$, where B is an observable and $F(t)$ a time-dependent classical field. Then the linear part of the resultant increment in the expectation value of an observable A at time t is

$$\Delta\langle A\rangle_t = \int_0^t \mathrm{d}t' L_{AB}(t-t')F(t'), \tag{6}$$

where

$$L_{AB}(t) = (\mathrm{i}/\hbar)\rho_c([A_t, B]_-), \tag{7}$$

† An earlier, limited version of this theorem is given by Nyquist's (1928) relation between electrical resistance and circuit noise.

with

$$A_t = e^{iHt/\hbar} A e^{-iHt/\hbar}. \tag{8}$$

On the other hand, the time-correlation function for the spontaneous equilibrium fluctuations of A and B in the unperturbed system is

$$C_{AB}(t) = \rho_c([A_t, B]_+), \tag{9}$$

and it is a straightforward matter to deduce† from eqns (1) and (7)–(9) that the Fourier transforms $\hat{L}_{AB}$, $\hat{C}_{AB}$ of L_{AB}, C_{AB}, respectively, are related by the formula

$$\hat{L}_{AB}(\omega) = \frac{i}{\hbar} \hat{C}_{AB}(\omega) \tanh \tfrac{1}{2}\beta\hbar\omega, \tag{10}$$

where

$$f(\omega) \equiv \int_{-\infty}^{\infty} dt\, e^{-i\omega t} f(t). \tag{11}$$

Equation (10) is usually referred to as the *fluctuation–dissipation* theorem, because C_{AB} represents spontaneous fluctuations in the unperturbed equilibrium state, while the response function L_{AB} governs the rate at which the system dissipates the energy supplied by the perturbation.† For example, if $F(t)$ is an electric field applied to a dielectric material and B is its total polarization, then the frequency–susceptibility $\chi(\omega)$ is $\int_{-\infty}^{\infty} dt\, e^{-i\omega t} L_{BB}(t)$, and so the rate of dissipation of field energy is $\frac{1}{2}\text{Im} \int d\omega \chi(\omega) |\hat{F}(\omega)|^2 = \frac{1}{2}\text{Im} \int d\omega \hat{L}_{BB}(\omega) |\hat{F}(\omega)|^2$.

By eqns (7) and (9), the fluctuation–dissipation formula (10) is the Fourier transform of the equation

$$\rho_c(A_t B) = \rho_c(B A_{t+i\hbar\beta}) \tag{12}$$

which is the basis of the Green function formulation of Statistical Thermodynamics by Martin and Schwinger (1959). Furthermore, as we shall prove in Appendix A, this formula completely characterizes ρ_c, in the sense that this is the only state satisfying the condition‡ that

$$\rho(A_t B) = \rho(B A_{t+i\hbar\beta}) \tag{13}$$

for all bounded observables A and B. This is customarily termed *the Kubo–Martin–Schwinger, or KMS, condition.*

Thus, the equilibrium states of finite systems are equivalently charac-

† See Kubo (1957).

‡ To be precise, eqn (13) should be interpreted as meaning that the function $f(t) \equiv \rho(BA_t)$ may be extended to a function $f(z)$, which is analytic in the interior of the strip $\text{Im}\, z \in [0, \hbar\beta]$ and continuous on its boundaries, and which satisfies the equation $f(t + i\hbar\beta) = \rho(A_t B)$ for real t.

terized by the thermodynamical stability and the KMS fluctuation–dissipation conditions. We shall now briefly review the argument, which we shall expound in the following sections of this chapter, by which these characterizations of equilibrium states are generalized to infinite systems.

Turning first to thermodynamical stability, we remark that infinite systems, unlike finite ones, admit a clear-cut distinction between stability at global and local levels. Here the essential idea, which will be precisely formulated in §3.4, is that a state is *globally stable* if it minimizes the free energy density of the entire system; while a state is *locally stable* if no modification of it, that is confined to a bounded spatial region, can lower the free energy of the system. The relationship between these two kinds of stability has been proved† to be quite simple. To be specific, the globally stable states are always locally stable, while the converse is valid for systems with suitable short-range forces. On the other hand, certain systems with long-range forces can support states that are locally, but not globally, stable: we shall argue in due course (Chapter 6) that these states are metastable.

As regards the KMS condition (13), this may be carried over to infinite systems,‡ within the terms of the algebraic description of the previous chapter, the observables A and B now being *local* ones. Correspondingly, the equivalence between this condition and thermodynamical stability carries through, at the local level,§ to infinite systems. Furthermore, the KMS condition as applied to an infinite system, has been proved¶ to characterize the states for which the system behaves as a *thermal reservoir*, in the sense that it drives any finite system, to which it is locally and weakly coupled, into equilibrium at temperature T. This reservoir property is precisely what is demanded by the Zeroth Law of Thermodynamics.

Thus, for an infinite system, the global stability and KMS conditions represent the requirements of the Second and Zeroth Laws, respectively. Furthermore, the former condition covers the latter one, since the KMS relation represents merely local stability. Hence, assuming that the equilibrium states are those that meet the demands of the Second and Zeroth Laws, we are led to the conclusion that they are completely characterized by the *global stability condition*. As we shall presently see, this characterization opens up the way for a theory of the phase structure of matter, since it emerges that an infinite system may support different globally stable states under the same thermodynamical conditions. These states may, for ex-

† See Sewell (1980).

‡ The importance of the KMS condition, in the context of the statistical mechanics of infinite systems, was first evinced by Haag, Hugenholtz, and Winnink (1967).

§ See Araki and Sewell (1977); Sewell (1977*a*, 1980).

¶ The essential idea here stems from the original treatment of the fluctuation–dissipation theorem by Callen and Welton (1951). It was subsequently developed, in a rigorous form, by Sewell (1974) for a limited class of models and by Kossakowski, Frigerio, Gorini, and Verri (1977) in a general context.

ample, be infinite volume limits of canonical states of finite systems with different boundary conditions.

We shall present the theory of equilibrium states, outlined above, as follows. Section 3.2 will be a preliminary mathematical section, dealing with the rudiments of the theory of convexity and continuity relevant to thermodynamical stability. In §3.3, we shall formulate the model on which our explicit treatment of infinite systems will be based. For simplicity, we shall take this to be a generic lattice model, Σ: discussion of the generalization of the theory to continuous systems will be left until the end of the chapter. In §3.4, we shall define the thermodynamic functionals of the states of the infinite system, Σ, and specify both the global and the local stability conditions in terms of these. In §3.5, we shall show that an infinite system may support different globally stable states under the same thermodynamical conditions. In particular, we shall show, by an adaptation of a classic argument of Peierls (1936), that the two-dimensional Ising model has two globally stable states, with equal and opposite polarizations, at sufficiently low temperatures. In §3.6, we shall examine the relationship between global and local stability and show, by an explicit treatment of a simple model, that certain systems with long-range forces can support states that are locally, but not globally, stable. In §3.7, we shall examine the significance of the KMS fluctuation–dissipation condition with regard to local stability and the Zeroth Law of Thermodynamics: in particular, we sketch a proof that this condition characterizes the states with the reservoir property, described above, corresponding to the Zeroth Law. In §3.8, we shall conclude, on the basis of the results of the previous Sections, that the equilibrium states of an infinite system are the globally stable ones; and that such a system may support several equilibrium states, corresponding to different phases, under the same thermodynamical conditions. We shall close §3.8 with a brief discussion of the generalisation of the theory of equilibrium states to continuous systems. The four Appendices will be devoted to proofs of mathematical assertions made in the main body of this chapter.

3.2. Mathematical preliminaries: convexity, continuity, and semicontinuity

We present here the rudiments of the theory of convex and of semicontinuous functions, as required for the formulation of thermodynamical stability conditions.

3.2.1. Convexity†

A region, K, of a vector, space, V, is termed *convex* if, for any two points v_1, v_2 of K and for $0<\lambda<1$, the point $\lambda v_1+(1-\lambda)v_2$ also lies in K; i.e. the

† For a general treatment of convexity, see for example, the book of Roberts and Varberg (1973).

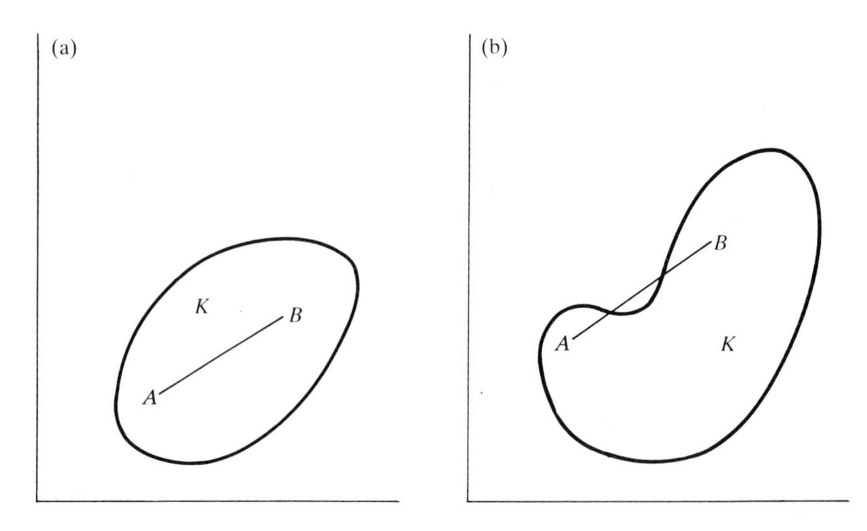

Fig. 1. (a) An oval-shaped region K is convex, as the line joining any two of its points, A and B, lies wholly in K. **(b)** Here the region K is not convex, as the line joining two of its points, A and B, does not lie wholly in K.

line joining v_1 to v_2 lies in K. Illustrations of convex and non-convex regions are given by Figs. 1(a) and 1(b) respectively.

A real-valued function $g(v)$ is termed *convex*, in a convex region K, if

$$g(\lambda v_1 + (1-\lambda)v_2) \leq \lambda g(v_1) + (1-\lambda)g(v_2) \tag{14}$$

for all v_1, v_2 in K and $0<\lambda<1$; and *strictly convex* if the strict inequality sign $<$ is applicable in (14) whenever $v_1 \neq v_2$. The definition (14) signifies that $g(v)$ is convex if its graph between two arbitrary points A and B lies below, or possibly coincides with, the chord AB (cf. Fig. 2). In the case where g is a differentiable function of a real variable v, this definition is equivalent to the condition that its derivative is a non-decreasing function of v.

Note. It follows easily by induction that the convexity condition (14) is equivalent to

$$g\left(\sum_g p_j v_j\right) \leq \sum_j p_j g(v_j)$$

for any $v_1, \ldots, v_n$ in K and any non-negative numbers $p_1, \ldots, p_n$, whose sum is unity. Thus one may characterize convexity by the property that

$$g(\langle v \rangle) \leq \langle g(v) \rangle, \tag{15}$$

where $\langle\ \rangle$ denotes the average for the probability distribution $(p_1, \ldots, p_n)$. Further, in the case of a strictly convex function $g(v)$, the strict inequality

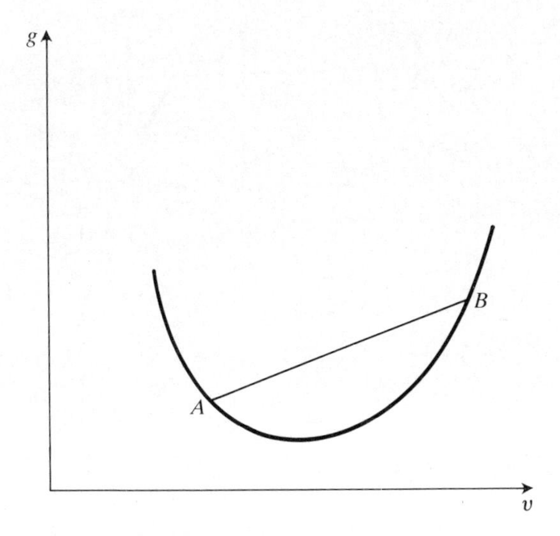

Fig. 2. Graph of a convex function g. Note that the chord AB lies above the graph between A and B.

$<$ is applicable in (15) except for the trivial cases where $v_1 = v_2 \ldots = v_n$ or all but one of the p's is zero.

Concave and affine functions. A function $g(v)$ is termed *concave*, over the convex region K, if $-g(v)$ is convex there, i.e. if the inequality in (14) goes the other way. On the other hand, a function $g(v)$ is termed *affine*, over the region K, if the equality sign is applicable in (14). Thus, an affine function is both convex and concave, in the sense of the standard definitions we have employed. One can put this in another way by saying that an affine function acts linearly on linear combinations, $\lambda v_1 + (1-\lambda)v_2$, of vectors v_1, v_2 in K, for values of λ between 0 and 1.

Some consequences of convexity. Let us now note the following restrictions imposed by the convexity of a function.

(1) A convex function cannot have two minima separated by a maximum (cf. Fig. 3(a)): nor, similarly, can it have a local non-absolute minimum.

(2) A convex function cannot have a discontinuity in the interior of its domain of definition (cf. Fig. 3(b)), though it might have a discontinuous derivative there (cf. Fig. 3(c)).

(3) A *strictly convex* function $g(v)$ cannot attain an absolute minimum value, g_0, say, at two different points, v_1 and v_2. For if it did so, the strict

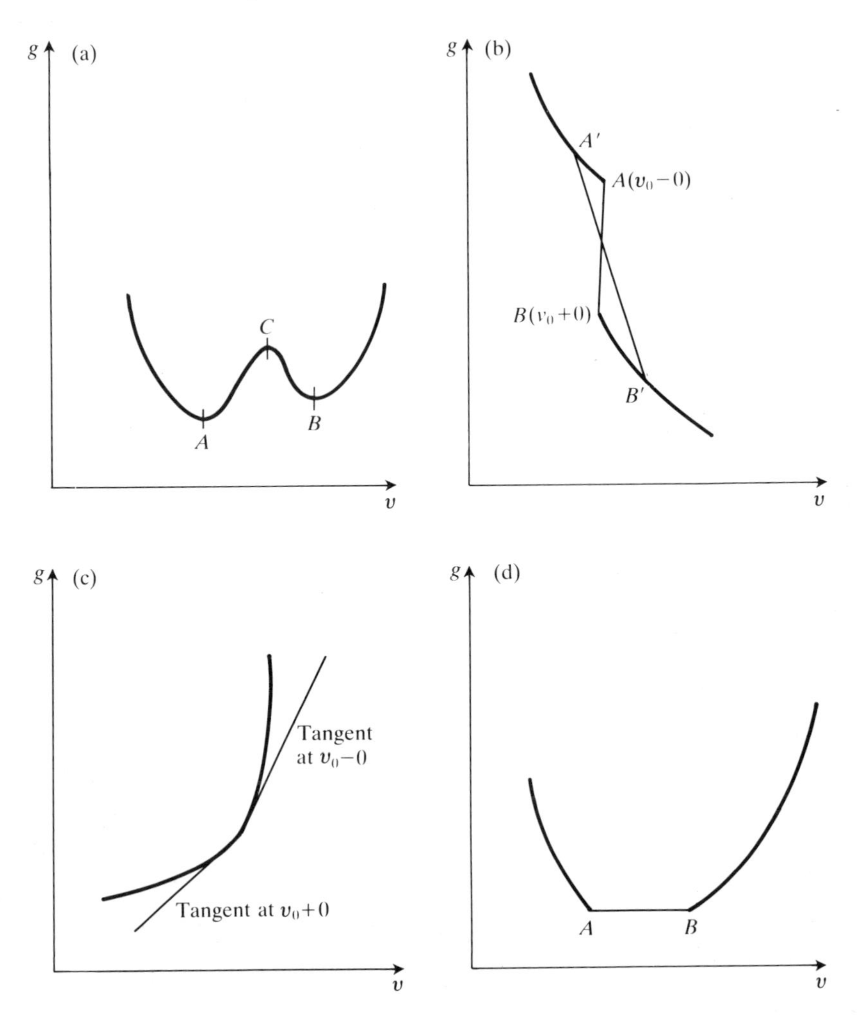

Fig. 3. (a) Function with minima at A and B separated by a maximum at C, is *not convex*, as C lies above the chord AB. **(b)** Function $g(v)$ with discontinuity at v_0 is *not convex*, since A lies above the chord $A'B'$. **(c)** Convex function with discontinuous derivative at v_0. **(d)** Convex function which is affine between A and B, and which attains its absolute minimum throughout the segment, AB, parallel to the v-axis.

convexity of g would imply that $g(\lambda v_1 + (1-\lambda)v_2) < g_0$, which would contradict the assumption that g_0 were the absolute minimum value of $g(v)$.

Note. This argument is not applicable to a function g that is affine, instead of strictly convex, over some domain Δ. For such a function could attain an absolute minimum over a range of points in Δ (cf. Fig. 3(d)).

3.2.2. *Note on convergent subsequences: the Bolzano–Weierstrass theorem*

As a preliminary to our considerations on continuity, we recall the classic Bolzano–Weierstrass theorem,† which tells us that *any bounded sequence of numbers* $\{x_n\}$ *has at least one convergent subsequence*.

A corollary to this theorem is that, if D is a closed,‡ bounded region of a Euclidean space Y, say, then *any sequence* $\{y_n\}$ *of points in* D *has a convergent subsequence*, *whose limit also lies in* D: this may be inferred by applying the Bolzano–Weierstrass theorem to each Cartesian coordinate separately.

A further simple corollary§ of the Bolzano–Weierstrass theorem is that, if $g(y)$ is a bounded, real-valued function of position over this region D, then *any sequence of points* $\{y_n\}$ *in* D *has a convergent subsequence* $\{y'_n\}$, *say*, *such that* $g(y'_n)$ *also converges as* $n \to \infty$.

Note. This result does not depend on whether or not $g(y)$ is continuous. The question of continuity pertains only to whether or not

$$\lim_{n\to\infty} g(y'_n) = g\,(\lim_{n\to\infty} y'_n).$$

3.2.3. *Semicontinuity*¶

A real-valued function $g(y)$ of position over a closed bounded region D of a Euclidean space Y is termed *upper semicontinuous* if, for any sequence $\{y_n\}$ of points in Y, such that $y_n \to \bar{y}$ and $g(y_n) \to \bar{g}$ as $n \to \infty$,

$$g(\bar{y}) \geq \bar{g};$$

and it is termed *lower semicontinuous* if the inequality goes the other way. Thus, full continuity is upper plus lower semicontinuity.

In general, semicontinuity properties are important for questions of whether a function attains an absolute maximum or minimum, as distinct from approaching an upper or lower bound arbitrarily closely. Such questions are obviously crucial to variational principles related to stability. We shall now demonstrate the relevance of semicontinuity to these questions by proving that, *if* $g(y)$ *is upper semicontinuous over the closed bounded region* D *of the Euclidean space* Y, *then* g *attains an absolute maximum value there*.

The proof runs as follows. If $\bar{g}$ is the least upper bound of $g(y)$ over the region D, then $g(y)$ may be taken to be arbitrarily close to $\bar{g}$ by a suitable

† See Whittaker and Watson (1946), p. 12.

‡ Recall that a closed region is one that contains the limit points of all its convergent sequences.

§ The proof of this is that $\{y_n\}$ has a convergent subsequence $\{y''_n\}$, and, by the Bolzano–Weierstrass theorem, this in turn has a subsequence $\{y'_n\}$ such that $g(y'_n)$ converges as $n \to \infty$.

¶ For a general treatment of semicontinuity, see, for example, Ch. 2, §7 of the book by Choquet (1966).

choice of y. This means that D contains a sequence $\{y_n\}$ such that $g(y_n) \to \bar{g}$ as $n \to \infty$; and by the Bolzano–Weierstrass theorem, we may take it that $\{y_n\}$ converges to a point $\bar{y}$, say, in D, since this can always be arranged by passing to a subsequence, if necessary. Thus, we have the situation that $y_n \to \bar{y}$ and $g(y_n) \to \bar{g}$ as $n \to \infty$. Since g is upper semicontinuous, this implies that $g(\bar{y}) \geqslant \bar{g}$. On the other hand, as $\bar{g}$ is an upper bound of $g(y)$ over D, the opposite inequality, $g(\bar{y}) \leqslant \bar{g}$, is also valid. Hence, $g(\bar{y}) = \bar{g}$, which means that the function $g(y)$ attains an absolute maximum at one point at least, namely $\bar{y}$, in D.

Similarly, *a lower semicontinuous function attains an absolute minimum over a closed bounded region of a Euclidean space*.

Note. These results concerning the attainment of maxima and minima have natural generalizations† to semicontinuous functions on a much wider class of spaces.

3.3. The model

We base our treatment of infinite systems on the model, Σ, of atoms or spins on a d-dimensional simple cubic lattice X, given by Example 1 of §2.4.1. We shall now provide fuller specifications of the model.

Thus, we first recall that the pure states and observables of an *isolated* atom of Σ are represented by the vectors and self-adjoint operators, respectively, in a finite-dimensional Hilbert space, $\mathcal{H}_0$; and, correspondingly, the observables of Σ that are localized at a site x of the lattice are represented by operators in a copy, $\mathcal{H}_x$, of $\mathcal{H}_0$. The algebra of observables, $\mathcal{A}(\Lambda)$, for a bounded region Λ of the system then consists of the operators in the Hilbert space, $\mathcal{H}(\Lambda)$, formed by the tensor product $\bigotimes_{x\in\Lambda} \mathcal{H}_x$, with the rule that, if $\Lambda \subset \Lambda'$, the elements A, of $\mathcal{A}(\Lambda)$ and $A \otimes I_{\Lambda\backslash\Lambda}$, of $\mathcal{A}(\Lambda')$, are identified with one another. The algebra $\mathcal{A}$ of bounded local observables of Σ is then the union of the local algebras $\mathcal{A}(\Lambda)$. This algebra is equipped with a group of automorphisms $\{\gamma(x)\}$ corresponding to space translations, and the local algebras of $\mathcal{A}(\Lambda)$ transform covariantly with respect to these translations, i.e.

$$\gamma(x)\mathcal{A}(\Lambda) = \mathcal{A}(\Lambda + x).$$

The *states* of Σ are the positive, normalized, linear functionals, ρ, on $\mathcal{A}$, such that the restriction of ρ to a bounded region Λ corresponds to a density matrix, ρ_Λ, in $\mathcal{H}(\Lambda)$, i.e.

$$\rho(A) = \mathrm{Tr}_\Lambda(\rho_\Lambda A) \quad \text{for } A \text{ in } \mathcal{A}(\Lambda). \tag{16}$$

A state ρ is termed *translationally invariant* if $\rho(\gamma(x)A) = \rho(A)$ for all local

† To be precise, an upper (resp. lower) semicontinuous function on a compact region of a topological space attains its absolute maximum (resp. minimum) there (see Choquet 1966, Ch. 2, §7).

observables A and lattice sites x. The definition we employ for *convergence*† of states is that a sequence $\{\rho_n\}$ converges to ρ if

$$\lim_{n\to\infty} \rho_n(A) = \rho(A) \quad \text{for all } A \text{ in } \mathscr{A}. \tag{17}$$

As we shall prove in Appendix B, the Bolzano–Weierstrass theorem may be extended to the states of Σ, in the sense that they enjoy the following property (P).

(P) *Any sequence* $\{\rho_n\}$ *of states of* Σ *has a convergent subsequence.*

The set of all states of Σ is termed *compact* by virtue of this property, whose usefulness arises in various constructions that require convergent sequences of states.

The *interactions* in Σ correspond to the family of local Hamiltonians, $H(\Lambda)$, for the bounded regions, Λ. Here, $H(\Lambda)$ is the observable given by the total energy of interaction between the particles in Λ. Thus, the energy of interaction between the atoms in disjoint regions Λ_1 and Λ_2 is

$$W(\Lambda_1, \Lambda_2) = H(\Lambda_1 \cup \Lambda_2) - H(\Lambda_1) - H(\Lambda_2) \tag{18}$$

We make the following assumptions concerning the interactions.

(a) They are translationally invariant, i.e.

$$\gamma(x)H(\Lambda) = H(\Lambda + x) \tag{19}$$

(b) The energy density is uniformly bounded, i.e.

$$\|H(\Lambda)\| < C\,|\Lambda| \tag{20}$$

for all bounded regions Λ, where C is a finite constant and $|\Lambda|$ is the number of sites in Λ.

(c) The interaction energy between the atoms in disjoint regions, Λ_1 and Λ_2, behaves as a 'surface effect' in the sense that

$$\|W(\Lambda_1, \Lambda_2)\| < \sigma(\Lambda_1) + \sigma(\Lambda_2), \tag{21}$$

where, for some positive ε,

$$\lim_{\Lambda\uparrow} \sigma(\Lambda)/|\Lambda|^{(1-\varepsilon)} = 0, \tag{22}$$

the limit being taken over some sequence of cubes or, more generally (cf. Fisher 1964), of regions sufficiently regular to ensure that the surface area of Λ increases as $|\Lambda|^{(d-1)/d}$, d being the dimensionality of the lattice. The inequality (21) signifies essentially that energy is an *extensive* variable.

We shall sometimes be concerned with interactions which are of *finite range* in the sense that, for any bounded region Λ and for Λ' containing Λ,

† This is the so-called W^*-convergence.

the value of $W(\Lambda, \Lambda'\backslash\Lambda)$ becomes independent of Λ' whenever the minimum Euclidean distance between the boundaries of Λ and Λ' exceeds some fixed distance, R. Thus, for example, the interactions in Heisenberg and Ising models with nearest neighbour couplings are of finite range.

The *dynamics* of Σ is given by the infinite volume limit of that of a finite system of particles of the given species, according to the scheme of §2.4.5. Thus, formally,† the time-translate of a local observable A is

$$A_t = \lim_{\Lambda\uparrow} \mathrm{e}^{\mathrm{i}H(\Lambda)t/\hbar} A \mathrm{e}^{-\mathrm{i}H(\Lambda)t/\hbar}. \tag{23}$$

A *symmetry* of Σ is defined as an automorphism α of the algebra which commutes with time translations, i.e.

$$\alpha A_t = (\alpha A)_t$$

The symmetry is termed *local* if α leaves the local Hamiltonians $H(\Lambda)$ invariant and maps each local algebra $\mathscr{A}(\Lambda)$ into itself. It follows from these definitions that the space-translational automorphisms $\{\gamma(x)\}$ form a group of symmetries of the system; while examples of local symmetries are provided by operations of spin reversals in the Heisenberg or Ising ferromagnets, or rotations in the Heisenberg model.

A state ρ is said to possess the symmetry α if $\rho(\alpha A) = \rho(A)$ for all local observables A.

3.4. Thermodynamical functionals and stability conditions

We shall base our formulation of thermodynamical stability conditions, at global and local levels, on two functionals of the states of Σ. The first of these will represent the free energy density of the entire system: the second will correspond to the increment in its free energy, due to strictly localized modifications of state. Both of these functionals will be defined as limiting forms of the corresponding ones for a finite system of particles of the given species.

3.4.1. *Local functionals*

Suppose that $\Sigma(\Lambda)$ is the finite system comprising the atoms in the bounded region Λ of the infinite lattice X. Then a state ρ of Σ will induce the state ρ_Λ on $\Sigma(\Lambda)$, and the resultant energy, entropy and free energy of this finite system will be given by

$$\hat{E}_\Lambda(\rho) = \mathrm{Tr}_\Lambda(\rho_\Lambda H(\Lambda)) \equiv \rho(H(\Lambda)) \tag{24}$$

$$\hat{S}_\Lambda(\rho) = -k\,\mathrm{Tr}_\Lambda(\rho_\Lambda \ln \rho_\Lambda), \tag{25}$$

† We recall from §2.4.5 that A_t, as given by (23), should be considered as a functional of the states.

and

$$\hat{F}_\Lambda(\rho) = \hat{E}_\Lambda(\rho) - T\hat{S}_\Lambda(\rho) \equiv \mathrm{Tr}_\Lambda(\rho_\Lambda H(\Lambda) + kT\rho_\Lambda \ln \rho_\Lambda), \tag{26}$$

respectively. We term the functionals $\hat{E}_\Lambda$, $\hat{S}_\Lambda$, and $\hat{F}_\Lambda$, defined by these formulae, *local*. These functionals enjoy the following key mathematical properties.†

(a) $\hat{E}_\Lambda$, $\hat{S}_\Lambda$, and $\hat{F}_\Lambda$ are all *continuous*, i.e. if $\rho_n \to 0$, then $\hat{E}_\Lambda(\rho_n) \to \hat{E}_\Lambda(\rho)$, etc. as $n \to \infty$. In fact, the continuity of $\hat{E}_\Lambda$ follows immediately from (24) and the boundedness of $H(\Lambda)$; while that of $\hat{S}_\Lambda$, and thence of $\hat{F}_\Lambda$, stems from the continuity of $x \ln x$, for real positive x, and the finite-dimensionality of $\mathscr{H}(\Lambda)$.

(b) $\hat{E}_\Lambda(\rho)/|\Lambda|$, $\hat{S}_\Lambda(\rho)/|\Lambda|$, and $\hat{F}_\Lambda(\rho)/|\Lambda|$ are all bounded, uniformly with respect to ρ and Λ. In fact, the boundedness of $\hat{E}_\Lambda(\rho)/|\Lambda|$ follows immediately from that of the energy density $H(\Lambda)/|\Lambda|$ (cf. (20)); while that of $\hat{S}_\Lambda(\rho)/|\Lambda|$, and thence of $\hat{F}_\Lambda(\rho)/|\Lambda|$, stems from the finite-dimensionality of the atomic Hilbert space $\mathscr{H}_0$.

(c) $\hat{E}_\Lambda$ is *affine*, S_Λ is *concave* and F_Λ is *convex*, i.e. (cf. §3.2.1)

$$\hat{E}_\Lambda(\lambda\rho_1 + (1-\lambda)\rho_2) = \lambda\hat{E}_\lambda(\rho_1) + (1-\lambda)\hat{E}_\Lambda(\rho_2)$$

$$\hat{S}_\Lambda(\lambda\rho_1 + (1-\lambda)\rho_2) \geqslant \lambda\hat{S}_\Lambda(\rho_1) + (1-\lambda)\hat{S}_\Lambda(\rho_2)$$

and

$$\hat{F}_\lambda(\lambda\rho_1 + (1-\lambda)\rho_2) \leqslant \lambda\hat{F}_\Lambda(\rho_1) + (1-\lambda)\hat{F}_\Lambda(\rho_2)$$

for $0<\lambda<1$. Here, the affine property of $\hat{E}_\Lambda$ follows immediately from (24); while the concavity of $\hat{S}_\Lambda$, and thus the convexity of $\hat{F}_\Lambda$, stems from the fact that $x \ln x$ is a concave function of the positive real variable x. We note that the concavity of $\hat{S}_\Lambda$ has a simple interpretation, namely that entropy is gained by mixing states of a finite system.

(d) $\hat{S}_\Lambda$ possesses the property that

$$\hat{S}_{\Lambda_1 \cup \Lambda_2}(\rho) \leqslant \hat{S}_{\Lambda_1}(\rho) + \hat{S}_{\Lambda_2}(\rho) \quad \text{for disjoint } \Lambda_1, \Lambda_2. \tag{27}$$

The local entropy functional $\hat{S}_\Lambda$ is termed *subadditive* by virtue of this property. In fact, one sees from the formula (27) that the subadditivity of entropy amounts to a generalized form of a classical information–theoretical result – namely, that the total information contained by a system is not less than the sum of the information contents of its parts.

An important, stronger version of the subadditivity property of $\hat{S}_\Lambda$,

† See Lanford and Robinson (1968*a*).

established by Lieb and Ruskai (1973), is that

$$\hat{S}_{\Lambda_1 \cup \Lambda_2}(\rho) + \hat{S}_{\Lambda_1 \cap \Lambda_2}(\rho) \leqslant \hat{S}_{\Lambda_1}(\rho) + \hat{S}_{\Lambda_2}(\rho) \quad \text{for } any \text{ bounded regions } \Lambda_1 \text{ and } \Lambda_2 \tag{28}$$

The local entropy is termed *strongly subadditive* by virtue of this property. As we shall presently see, the subadditivity properties of entropy enable one to prove the existence of certain infinite volume limits of the thermodynamic functionals.

3.4.2. Global functionals

Suppose now that ρ is a translationally invariant state of Σ. Then it follows from standard arguments, which we shall present in Appendix (C), that the 'surface condition' (21) on the interactions implies the convergence of the energy density $\hat{E}_\Lambda(\rho)/|\Lambda|$, as Λ increases to infinity; and that the subadditivity of entropy implies the convergence of $\hat{S}_\Lambda(\rho)/|\Lambda|$ in this limit. Hence, we may define the *global energy, entropy and free energy density functionals* of the translationally invariant states to be

$$\hat{e}(\rho) = \lim_{\Lambda\uparrow} \hat{E}_\Lambda(\rho)/|\Lambda|, \tag{29}$$

$$\hat{s}(\rho) = \lim_{\Lambda\uparrow} \hat{S}_\Lambda(\rho)/|\Lambda|, \tag{30}$$

and

$$\hat{f}(\rho) = \hat{e}(\rho) - T\hat{s}(\rho) = \lim_{\Lambda\uparrow} \hat{F}_\Lambda(\rho)/|\Lambda|, \tag{31}$$

respectively. It follows immediately from these formulae, together with the uniform boundedness of $\hat{E}_\lambda/|\Lambda|$, $\hat{S}_\Lambda/|\Lambda|$, and $\hat{F}_\Lambda/|\Lambda|$ that *the functionals* $\hat{e}$, $\hat{s}$, *and* $\hat{f}$ *are all bounded*. These functionals also possess the following key properties, whose proofs† we shall sketch in Appendix C.

(1) $\hat{e}$, $\hat{s}$, *and* $\hat{f}$ *are all affine*, i.e. $\hat{e}(\lambda\rho_1 + (1-\lambda)\rho_2) = \lambda\hat{e}(\rho_1) + (1-\lambda)\hat{e}(\rho_2)$, etc. Here, the affine property of $\hat{e}$ follows immediately from that of $\hat{E}_\Lambda$. On the other hand, even though $\hat{S}_\Lambda/|\Lambda|$ is strictly concave, it turns out that its limit, $\hat{s}$, is affine. In other words, although entropy is gained by mixing states of a finite system, the gain in entropy density vanishes in the infinite volume limit.

(2) $\hat{e}$ *is continuous*, $\hat{s}$ *is upper semicontinuous and* $\hat{f}$ *is lower semicontinuous*, i.e. if $\rho_n \to \rho$, and $\hat{e}(\rho_n)$, $\hat{s}(\rho_n)$ and $\hat{f}(\rho_n) \to \bar{e}$, $\bar{s}$, and $\bar{f}$, respectively, as $n \to \infty$, then $\bar{e} = \hat{e}(\rho)$, $\bar{s} \leqslant \hat{s}(\rho)$, and $\bar{f} \geqslant \hat{f}(\rho)$.

These properties stem from the uniform boundedness of the energy density $H(\Lambda)/|\Lambda|$ and the sub-additivity of entropy.

† See Lanford and Robinson (1968*a*) for detailed proofs.

The boundedness and lower semicontinuity of $\hat{f}$ are directly relevant to the stability problem of whether this functional attains an absolute minimum value. For, in view of the compactness property (P)† of the states of Σ we may employ the argument of §3.2.3 concerning the extremal properties of bounded semicontinuous functions to prove that $\hat{f}$ *attains an absolute minimum value, for at least one translationally invariant state*. Furthermore, for reasons we shall presently outline,‡ *the minimum value of* $\hat{f}$ *is equal to the infinite volume limit of the canonical free energy density of a finite system of the given species at temperature* T, i.e.

$$\min \hat{f}(\rho) = \phi(T), \tag{32}$$

where

$$\phi(T) = \lim_{\Lambda\uparrow} \Phi_\Lambda(T)/|\Lambda|, \tag{33}$$

and

$$\Phi_\Lambda(T) = -kT \ln \mathrm{Tr}_\Lambda \exp - (\beta H(\Lambda)). \tag{34}$$

Thus, (32) *is the analogue of the variational principle given by* (15) *for finite systems*.

Its proof runs as follows. On applying the variational principle (15) to the finite system, $\Sigma(\Lambda)$, comprising the atoms in the bounded region Λ, we see that

$$\hat{F}_\Lambda(\rho) \geqslant \Phi_\Lambda(T)$$

for any state ρ of Σ. Thus, dividing this formula by $|\Lambda|$ and passing to the infinite volume limit, we obtain the inequality $\hat{f}(\rho) \geqslant \phi(T)$, for any translationally invariant state ρ, from which it follows that

$$\min \hat{f}(\rho) \geqslant \phi(T). \tag{35}$$

Therefore, in order to establish that the equality sign is operative here, in accordance with (32), it suffices to construct a translationally invariant state, $\bar{\rho}$, whose free energy density is $\phi(T)$.

To this end, we construct a translationally invariant infinite volume limit of a canonical state by the following procedure. We resolve the lattice X into a disjoint set of copies $\{\Lambda_J\}$ of a cube Λ_0 of side L and centre O. We then define ρ_{Λ_J} to be the canonical state of the system $\Sigma(\Lambda_J)$ of atoms in Λ_J, as given by the density matrix $\exp(-\beta H(\Lambda_J))/\mathrm{Tr}(\mathrm{idem})$; and we define ρ_L to be the state of Σ which reduces to ρ_{Λ_J} in the cube Λ_J, and which carries no correlations between the different cubes, i.e.

$$\rho_L = \bigotimes_J \rho_{\Lambda_J}. \tag{36}$$

† Recall (cf. §3.2.2) that this is the property whereby any sequence of states of Σ has a convergent subsequence.

‡ See Lanford and Robinson (1968*a*) for a detailed proof.

In order to convert this state into a translationally invariant one, we average it over Λ_0, thereby obtaining

$$\bar{\rho}_L = |\Lambda_0|^{-1} \sum_{x \in \Lambda_0} \rho_{L,x}, \tag{37}$$

where ρ_x is the space translate of ρ by x, i.e. $\rho_x(A) \equiv \rho(\gamma(x)A)$. By the compactness property (P) of the states of Σ, $\bar{\rho}_L$ converges to a limit as L increases to infinity over some sequence of values. This limit,

$$\bar{\rho} = \lim_{L \to \infty} \bar{\rho}_L, \tag{38}$$

is then a space-averaged infinite volume canonical state. One may check† that it follows from the lower semicontinuity of $\hat{f}$, together with the convexity of the local free energy functional $\hat{F}_\Lambda$ and the surface property (21) of the interactions that

$$\hat{f}(\bar{\rho}) \leq \lim_{\Lambda_0 \uparrow} |\Lambda_0|^{-1} \Phi_{\Lambda_0}(T) \equiv \phi(T) \quad \text{(by (33))}.$$

Therefore, as the minimum value of $\hat{f}$ is not less than $\phi(T)$, by (35), it follows that

$$f(\bar{\rho}) = \phi(T) \tag{39}$$

as required.

Note. There is no claim here that $\bar{\rho}$ is the only state that minimizes $\hat{f}$.

3.4.3. *Incremental functionals*

Suppose that ρ and ρ' are states of Σ, which coincide outside some bounded region, Λ_0 say, i.e. $\rho'_{\Lambda_1} = \rho_{\Lambda_1}$ if Λ_0 and Λ_1 are disjoint. ρ' may then be regarded as a state obtained by a *localized modification* of ρ. We seek to define the increments in energy, entropy and free energy of the system, due to this modification, as limits of the resultant changes in $\hat{E}_\Lambda$, $\hat{S}_\Lambda$, and $\hat{F}_\Lambda$, as Λ increases to infinity. Now, since ρ' and ρ coincide outside Λ_0.

$$\begin{aligned}\hat{E}_\Lambda(\rho') - \hat{E}_\Lambda(\rho) &= (\hat{E}_\Lambda(\rho') - \hat{E}_{\Lambda \backslash \Lambda_0}(\rho')) - (\hat{E}_\Lambda(\rho) - \hat{E}_{\Lambda \backslash \Lambda_0}(\rho)) \\ &= \hat{E}_{\Lambda_0}(\rho') - \hat{E}_{\Lambda_0}(\rho) + \rho'(W(\Lambda_0, \Lambda \backslash \Lambda_0)) - \rho(W(\Lambda_0, \Lambda \backslash \Lambda_0)), \\ &\qquad \text{(by (18 and (24))}.\end{aligned}$$

Hence, $\hat{E}_\Lambda(\rho') - \hat{E}_\Lambda(\rho)$ converges, as Λ increases to infinity, if the difference between the expectation values of $W(\Lambda_0, \Lambda \backslash \Lambda_0)$ for the states ρ and ρ' converges in this limit. Under this condition, which is fulfilled by systems with finite range interactions as well as by a large class of others, we may define the *incremental energy*

$$\Delta \hat{E}(\rho \mid \rho') = \lim_{\Lambda \uparrow} (\hat{E}_\Lambda(\rho') - \hat{E}_\Lambda(\rho)). \tag{40}$$

† See Lanford and Robinson (1968*a*).

On the other hand, the change in $\hat{S}_\Lambda$ due to the transition from ρ to ρ' is

$$\hat{S}_\Lambda(\rho') - \hat{S}_\Lambda(\rho) = (\hat{S}_\Lambda(\rho') - \hat{S}_{\Lambda\backslash\Lambda_0}(\rho')) - (\hat{S}_\Lambda(\rho) - \hat{S}_{\Lambda\backslash\Lambda_0}(\rho)).$$

Now, a consequence† of the strong subadditivity of entropy is that, for any state ρ, $\hat{S}_\Lambda(\rho) - \hat{S}_{\Lambda\backslash\Lambda_0}(\rho)$ converges to a finite limit as Λ increases to infinity; the argument is that the strong subadditivity ensures that this entropy difference decreases monotonically with Λ, and that the finite-dimensionality of $\mathscr{H}(\Lambda_0)$ ensures that it has a finite lower bound. Thus, by the above equation for $(\hat{S}_\Lambda(\rho') - \hat{S}_\Lambda(\rho))$, this local entropy change converges to a finite limit $\Delta\hat{S}(\rho \mid \rho')$, as Λ increases to infinity, i.e.

$$\Delta\hat{S}(\rho \mid \rho') = \lim_{\Lambda\uparrow} (\hat{S}_\Lambda(\rho') - \hat{S}_\Lambda(\rho)). \tag{41}$$

We term $\Delta\hat{S}(\rho \mid \rho')$ the *incremental entropy* for the transition from ρ to ρ' and define the *incremental free energy* for that transition to be

$$\Delta\hat{F}(\rho \mid \rho') = \Delta\hat{E}(\rho \mid \rho') - T\Delta\hat{S}(\rho \mid \rho') = \lim_{\Lambda\uparrow} (\hat{F}_\Lambda(\rho') - \hat{F}_\Lambda(\rho)). \tag{42}$$

3.4.4. *Thermodynamical stability conditions*

We are now in a position to specify thermodynamical stability conditions for states of the infinite system in terms of the global and the incremental functionals we have just defined.

We term a translationally invariant state ρ of Σ *globally thermodynamically stable* (*GTS*) if it minimizes the free energy density functional $\hat{f}$, i.e. by (32), if

$$\hat{f}(\rho) = \phi(T). \tag{43}$$

Thus, for example, the state $\bar{\rho}$ constructed in §3.4.2 is GTS (cf. eqn (39)).

On the other hand, we term a state ρ of Σ *locally thermodynamically stable* (*LTS*) if no localized modification of it can lead to a reduction of the free energy of the system, i.e. if

$$\Delta\hat{F}(\rho \mid \rho') \geqslant 0 \tag{44}$$

for any state ρ' which coincides with ρ outside some bounded region.

We shall specify the relationship between the local and global stability conditions in due course (§3.6).

Note on translational invariance. The global stability condition (43), though ostensibly devised for fully translationally invariant states, is also applicable to states carrying the translational symmetry of any sublattice of X. It can therefore accommodate a description of, say, antiferromagnetic states. As regards the local stability condition (44), this does not involve any translational symmetry.

† See Lemma 2.4(1) of the article by Araki and Sewell (1977).

3.5. Global stability, macroscopic degeneracy, and symmetry breakdown

We shall now establish an important difference between finite and infinite systems, namely that, although a finite system possesses just one equilibrium state at temperature T, an infinite one can support several GTS states at the same temperature. A key to this difference is that the free energy density functional, $\hat{f}(\rho)$, of an infinite system is affine, whereas the free energy, $\hat{F}(\rho)$, of a finite one is strictly convex.† For, as noted at the end of §3.2.1, an affine function, unlikely a strictly convex one, can attain an absolute minimum value at more than one point; and this implies that, while the strict convexity of $\hat{F}(\rho)$ permits no more than one equilibrium state of a finite system at any given temperature, the affine property of $\hat{f}(\rho)$ leaves open the possibility that an infinite system might support several different globally stable states at the same temperature. In fact, this possibility has been proved to be realized in various tractable models.‡ For example, as we shall presently see, the two-dimensional Ising model possesses two different globally stable states, with oppositely directed polarizations, at any sufficiently low temperature.

In general, we shall refer to the existence of several GTS states at the same temperature as *macroscopic degeneracy*. The fact that an infinite system can exhibit such a degeneracy is of crucial importance for the theory of *phase transitions*, since it may be naturally interpreted as signifying that different phases of such a system can coexist under the same thermodynamical conditions.

Since macroscopic degeneracy can occur, we shall generally be concerned with a set of GTS states, rather than just one such state, at a given temperature. We remark here that it follows from the affine property of the free energy functional $\hat{f}$ that, if ρ and ρ' satisfy the GTS condition (43), then so too does $\lambda\rho + (1-\lambda)\rho'$ for any λ between 0 and 1. In other words, *mixtures of GTS states are themselves GTS*; or, in mathematical terms, the set of globally stable states at temperature T is *convex*.

A phenomenon closely related to macroscopic degeneracy is that of *symmetry breakdown*. This may be described as follows. Suppose that α is a local symmetry automorphism of the system, as defined in §3.3. Then the transformation $\rho \to \rho_\alpha$ of the states, induced by this automorphism is given by

$$\rho_\alpha(A) = \rho(\alpha A),$$

and it follows easily from our definition of the free energy density function $\hat{f}$

† The strict convexity of $\hat{F}(\rho)$ may easily be derived from that of the function $x \ln x$ of the positive real variable x, by the methods of Lanford and Robinson (1968*a*).

‡ Apart from the Ising model, other examples are provided by the two-dimensional anisotropic Heisenberg ferromagnet (see Ginibre 1969; Robinson 1969) and by various three-dimensional quantum lattice models (see Dyson, Lieb, and Simon 1978).

that

$$\hat{f}(\rho_\alpha) = \hat{f}(\rho).$$

In other words, $\hat{f}$ possesses the symmetry associated with α. From this it follows that if Δ, the set of GTS states, consists of only one element, $\bar{\rho}$ say, then $\bar{\rho}_\alpha = \bar{\rho}$, which means that this state possesses the symmetry α. On the other hand, in the case of a macroscopic degeneracy, it can happen that, while the whole set Δ is invariant with respect to the transformation $\rho \to \rho_\alpha$, certain individual elements of Δ are not. In this case, one says that there is a *spontaneous symmetry breakdown*. This is a phenomenon that occurs in many phase transitions. We shall now present a concrete example of a macroscopic degeneracy with symmetry breakdown.

3.5.1. Macroscopic degeneracy and symmetry breakdown in the two-dimensional Ising model

The two-dimensional Ising model is a system of spins, located at the sites of a square lattice X. For each point x of the lattice, the spin observable is a scalar,† $\sigma(x)$, possessing two eigenstates, with eigenvalues 1 and -1. The interactions between the spins are confined to nearest neighbours, the energy of interaction between spins at neighbouring sites x and x' being $-J\sigma(x)\sigma(x')$, where J is a positive constant. The local Hamiltonian corresponding to the energy of interaction between the spins in a bounded region Λ is therefore

$$H(\Lambda) = -J \sum_{\substack{x,x' \\ \in\Lambda}}{}' \sigma(x)\sigma(x'), \tag{45}$$

the prime over Σ indicating that summation is confined to nearest neighbouring sites.

We shall show that, at sufficiently low temperatures, this model supports two globally stable states, with equal and opposite polarizations, and thus exhibits macroscopic degeneracy with breakdown of the symmetry corresponding to spin reversals, $\sigma \to -\sigma$. The proof will be achieved in two main stages. In the first, we shall invoke a classic argument,‡ due to Peierls (1936) which shows that, at sufficiently low temperatures, a *finite* version of the model, with suitable boundary conditions, has an equilibrium polarization which survives the infinite volume limit. In the second stage, we shall use this result to construct polarized GTS states of the infinite system by certain limiting procedures.

The Peierls argument for a finite Ising model. We start by considering a finite Ising model, occupying a square region Λ of the lattice X, subject to the condition that the sites outside and neighbouring to the boundary of Λ are

† Equivalently, $\sigma(x)$ may be considered to be the z-component, say, of a Pauli spin.
‡ The rigorous form of this argument was provided by Griffiths (1964).

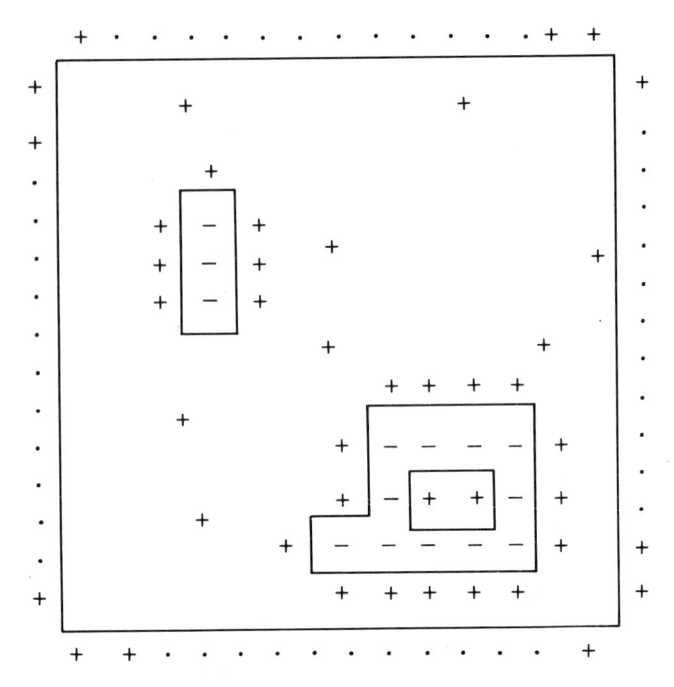

Fig. 4. The square region Λ with positive boundary conditions, and internal contours separating positive from negative spins.

occupied by spins whose values are fixed $+1$ (cf. Fig. 4): we refer to this as the *positive boundary condition*. Under this condition, the Hamiltonian for the system, $\Sigma(\Lambda)$, includes a contribution due to the interaction of these fixed positive spins with those in the 'surface layer' $\partial\Lambda$, consisting of the sites in Λ that have nearest neighbours outside that region. Thus, the total Hamiltonian is

$$H^{(+)}(\Lambda) = H(\Lambda) + \delta H(\Lambda), \tag{46}$$

where

$$\delta H(\Lambda) = -J \sum_{x \in \partial\Lambda} \sigma(x). \tag{47}$$

Correspondingly, the equilibrium state at temperature T is given by the density matrix

$$\rho_\Lambda^{(+)} = \mathrm{e}^{-\beta H^{(+)}(\Lambda)}/\mathrm{Tr}_\Lambda(\mathrm{idem}), \tag{48}$$

and the expectation value of the spin at x for this state is

$$\langle \sigma(x) \rangle_\Lambda^{(+)} = \mathrm{Tr}_\Lambda(\rho_\Lambda^{(+)} \sigma(x)) \tag{49}$$

We aim to show that, at sufficiently low temperatures, this value exceeds some strictly positive quantity m, which depends neither on the position of x in Λ nor on the size of Λ, i.e.

$$\langle\sigma(x)\rangle_{\Lambda}^{(+)} > m > 0, \quad \text{independently of } x\ (\in\Lambda) \text{ and of } \Lambda. \tag{50}$$

Peierls's argument leading to this result is based on the observation that each configuration of spins, with values ± 1, corresponds to a set of polygonal contours in Λ, which separate the regions of positive and negative spins (cf. Fig. 4): these contours may be taken to pass through the points midway between pairs of neighbouring sites with opposite spins. In order to express the energy for a spin configuration in terms of the associated Peierls contours, we note that, on the one hand, the energy required to reverse one of a pair of initially parallel, neighbouring spins is $2J$; while, on the other hand, the total number of neighbouring pairs of antiparallel spins in a configuration corresponding to a set of contours $\gamma_1, \ldots, \gamma_n$ is equal to the sum of their lengths, $\sum_1^n |\gamma_j|$, in units of the lattice spacing. Hence, for the positive boundary condition, the energy difference between this configuration and that where all the spins are positive is

$$E(\gamma_1, \ldots, \gamma_n) = 2J \sum_1^n |\gamma_j|. \tag{51}$$

From this formula, we may obtain a simple bound on the equilibrium probability that a polygon, γ, say, is a Peierls contour separating regions of positive and negative spins. For, if C and C' are configurations corresponding to sets of contours $(\gamma, \gamma_2, \ldots, \gamma_n)$ and $(\gamma_2, \ldots, \gamma_n)$, respectively, then it follows from (51) that the ratio of the equilibrium probabilities of C and C' is $e^{-2\beta J|\gamma|}$. Hence, in general, the total equilibrium probability, $P(\gamma)$, that γ is a Peierls contour, cannot exceed $e^{-2\beta J|\gamma|}$, i.e.

$$P(\gamma) \leqslant \exp(-2\beta J|\gamma|). \tag{52}$$

We can now use this inequality to obtain a bound on the probability that the spin, $\sigma(x)$, at an arbitrary site x, in Λ, is negative. For, in order that $\sigma(x)$ be negative, it is necessary (though not sufficient) that x is surrounded by at least one Peierls contour. Hence, the total probability, $P^{(-)}$, that the spin at x be negative is majorized by the sum of probabilities, $P(\gamma)$, for all the contours surrounding x,

$$P^{(-)} \leqslant \sum_{\gamma} P(\gamma) \leqslant \sum_{\gamma} \exp(-2\beta J|\gamma|), \quad \text{by (52)}.$$

Evidently, this inequality remains valid if we extend the sum on its RHS to all contours in the lattice X, not merely in Λ, that surround x. Hence, denoting by $N(l)$ the total number of such contours whose lengths are l, we see that

$$P^{(-)} \leqslant \sum_4^{\infty} N(l) e^{-2\beta J l}, \tag{53}$$

the lowest value of l being 4, as that is the minimum length of a closed polygonal contour. In order to obtain an estimate† for $N(l)$, and thus for $P^{(-)}$, we note the following simple facts.

(a) Any contour of length l consists of l unit links. Since a link has no more than four directions available to it (even ignoring constraints imposed by the requirement that the contour be polygonal), this implies that the number of different contours of length l passing through a point y, say, cannot exceed 4^l.

(b) The number of points y that can belong to contours of length l, which surround x, is less than $4l^2$, the area of the square of side $2l$ and centred at x.

It follows easily from (a) and (b) that

$$N(l) < 4l^2 \,.\, 4^l \equiv l^2 \,.\, 4^{l+1},$$

and therefore, by (53), that

$$P^{(-)} < \sum_4^{\infty} l^2 \,.\, 4^{(l+1)}\, \mathrm{e}^{-2\beta J l}$$

Since the RHS of this inequality converges for $\beta J > \ln 2$, and tends to zero as $\beta \to \infty$, i.e. as $T \to 0$, it follows that, given any number m between 0 and 1, there is some temperature, T_0, such that

$$P^{(-)} < \tfrac{1}{2}(1-m) \quad \text{for} \quad T < T_0$$

Consequently, the probability, $P^{(+)}$, that the spin at x is $+1$, satisfies the inequality

$$P^{(+)} > \tfrac{1}{2}(1+m) \quad \text{for} \quad T < T_0,$$

and therefore

$$\langle \sigma(x) \rangle_\Lambda^{(+)} \equiv P^{(+)} - P^{(-)} > m \quad \text{for} \quad T < T_0,$$

which proves the validity of the formula (50).

Note on the canonical free energy density. The result we have just obtained signifies that the positive boundary condition induces a polarization which survives the infinite volume limit. However, as we shall now show, *the boundary condition has no effect on the free energy density of the system* in that limit. To prove this point, we note that the free energies of the finite system $\Sigma(\Lambda)$, under positive and free boundary conditions, are

$$\Phi_\Lambda^{(+)}(T) = -kT \ln \mathrm{Tr}\, \mathrm{e}^{-\beta H^{(+)}(\Lambda)} \equiv \min_{\rho_\Lambda} \mathrm{Tr}[\rho_\Lambda H^{(+)}(\Lambda) + kT \rho_\Lambda \ln \rho_\Lambda] \quad (54)$$

† The estimate given here is very generous! For a better, i.e. more restrictive, estimate, see Griffiths (1964).

and

$$\Phi_{\Lambda}(T) = -kT \ln \mathrm{Tr}\, \mathrm{e}^{-\beta H(\Lambda)} \equiv \min_{\rho_{\Lambda}} \mathrm{Tr}[\rho_{\Lambda} H(\Lambda) + kT\rho_{\Lambda} \ln \rho_{\Lambda}],$$

respectively. Since, by (46), $H^{(+)}(\Lambda) = H(\Lambda) + \delta H(\Lambda)$, the Trace functionals of ρ_{Λ} on the RHS's of these formulae cannot differ by more than $\|\delta H(\Lambda)\|$. The same is therefore true of their minima, so that

$$|\Phi_{\Lambda}^{(+)}(T) - \Phi_{\Lambda}(T)|/|\Lambda| \leqslant \|\delta H(\Lambda)\|/|\Lambda|.$$

Moreover, it follows from the definition (47) of $\delta H(\Lambda)$ as a 'surface interaction' that the right-hand side of this inequality tends to zero as Λ increases to infinity. Hence, so too does the difference between $\Phi_{\Lambda}^{(+)}(T)/|\Lambda|$ and $\Phi_{\Lambda}(T)/|\Lambda|$. Therefore, by (33),

$$\lim_{\Lambda\uparrow} \Phi_{\Lambda}^{(+)}(T)/|\Lambda| = \phi(T), \tag{55}$$

which confirms that the free energy density is independent of the boundary condition in the infinite volume limit.

Polarized GTS state of the infinite Ising model. We now utilize the above results, (50) and (55), to construct a polarized GTS state of the infinite Ising model. This construction is analogous to that of the GTS state $\bar{\rho}$ in §3.4.1, though based on the finite volume canonical states $\rho_{\Lambda}^{(+)}$ with positive, rather than free, boundary conditions.

Thus, we resolve the lattice X into equal squares $\{\Lambda_J\}$ of side L, and define $\rho_{\Lambda_J}^{(+)}$ to be the canonical state of the finite spin system in Λ_J, with positive boundary conditions, as given by eqn (48). We define $\rho_L^{(+)}$ to be the state of the infinite systems which reduces to $\rho_{\Lambda_J}^{(+)}$ in each Λ_J and carries no correlations between different Λ_J's, i.e.

$$\rho_L^{(+)} = \bigotimes_J \rho_{\Lambda_J}^{(+)}$$

We then define $\bar{\rho}_L^{(+)}$ to be the translationally invariant state obtained by space-averaging $\rho_L^{(+)}$ over a cell Λ_0, say, i.e.

$$\bar{\rho}_L^{(+)} = |\Lambda_0|^{-1} \sum_{x\in\Lambda_0} \rho_{L,x}^{(+)},$$

where $\rho_{L,x}^{(+)}$ is the space translate of $\rho_L^{(+)}$ by x $(\rho_{L,x}^{(+)}(A) \equiv \rho_L^{(+)}(\gamma(x)A))$. Finally, we define $\bar{\rho}^{(+)}$ to be the limit of $\bar{\rho}_L^{(+)}$ as L tends to infinity over some sequence of lengths, i.e.

$$\bar{\rho}^{(+)} = \lim_{L\to\infty} \bar{\rho}_L^{(+)} \tag{56}$$

It now follows easily from this definition that $\bar{\rho}^{(+)}$ inherits the polarization of $\rho_{\Lambda}^{(+)}$, as expressed by (50), i.e.

$$\bar{\rho}^{(+)}(\sigma(x)) > m > 0 \tag{57}$$

Furthermore, the argument described in §3.4.1 to establish the GTS property of $\bar{\rho}$ can easily be adapted to prove that $\bar{\rho}^{(+)}$ is GTS. Specifically, it follows on applying that argument to $\bar{\rho}^{(+)}$ that

$$\hat{f}(\bar{\rho}^{(+)}) \leqslant \lim_{\Lambda\uparrow} |\Lambda|^{-1}\Phi_{\Lambda}^{(+)}(T),$$

and therefore, by (55), that

$$\hat{f}(\bar{\rho}^{(+)}) \leqslant \phi(T).$$

Since $\phi(T)$ is the minimum value of the functional $\hat{f}$, this means that the equality sign is operative here and that $\bar{\rho}^{(+)}$ is globally stable.

Thus, the model has a globally stable, positively polarized state $\bar{\rho}^{(+)}$ when the temperature is sufficiently low.

Macroscopic degeneracy and symmetry breakdown. Similarly, it also has then a globally stable, negatively polarized state, $\bar{\rho}^{(-)}$, given by a space-averaged infinite volume limit of a finite Ising model with negative spins placed in the layer just outside its boundary. In other words, the system has oppositely polarized states $\bar{\rho}^{(\pm)}$, corresponding to positive and negative boundary conditions. Thus, the system exhibits a macroscopic degeneracy with breakdown of the symmetry corresponding to spin reversals $\sigma \rightarrow -\sigma$.

3.6. Relationship between local and global stability

One might anticipate that global stability always implies local stability, and this has been proved† to be the case. The validity of the converse, however, depends essentially on the range of the interactions; and in fact it has been established‡ that

(a) translationally invariant LTS states of systems with finite-range interactions are GTS; while

(b) certain systems with long-range forces can support translationally invariant LTS states that are not GTS.

We shall presently confirm (b) by a simple example, and leave the proof of (a) until Appendix D. It will be seen that the key to these results is that the LTS and GTS conditions reduce to the same thing unless the forces are of sufficiently long range to produce an interaction between the particles inside and outside a 'large' bounded region, which corresponds to a *volume effect*, not merely a surface effect. As the following example shows, this can happen even when the interactions between the particles in two bounded regions satisfy the surface condition (21).

† See Araki and Sewell (1977) and Sewell (1980). We remark that the proof that GTS implies LTS is not trivial, since the global stability condition concerns the positivity of increments in the free energy *density* of the system, while that of local stability concerns increments in the free energy itself. In fact, the proof has been accomplished only via the KMS condition.

‡ See Araki and Sewell (1977) and Sewell (1977*b*, 1980).

3.6.1. Model with states that are locally, but not globally, stable

This is a system, Σ, of identical atoms on a one-dimensional lattice, X, that interact via many-body forces favouring the formation of 'clusters' of particles in certain atomic states: these forces are similar to those of Fisher's (1967*a*) model of condensation.

To formulate the present model, we resolve the atomic Hilbert space $\mathcal{H}_0$ into subspaces $\mathcal{H}_0^{(0)}$ and $\mathcal{H}_0^{(1)}$. Correspondingly, denoting the sites of the lattice X by the integers, we resolve the copy, $\mathcal{H}_n$, of $\mathcal{H}_0$ at the site n into subspaces $\mathcal{H}_n^{(0)}$ and $\mathcal{H}_n^{(1)}$. We denote the projection operator for $\mathcal{H}_n^{(0)}$ by $P^{(0)}(n)$. In the standard way, we define $\mathcal{H}(\Lambda) = \bigotimes_{n\in\Lambda} \mathcal{H}_n$ to be the state space for the atoms in a bounded region Λ, and we define $\mathcal{H}^{(0)}(\Lambda) = \bigotimes_{n\in\Lambda} \mathcal{H}_n^{(0)}$ to be its subspace based on $\mathcal{H}_0^{(0)}$-class spaces.

We assume that the interactions in Σ consist of (a) short-range ones that operate between atoms in $\mathcal{H}_0^{(0)}$-class states, and (b) many-body interactions that favour the formation of clusters of atoms in those states. Specifically we assume that the local Hamiltonian is of the form

$$H(\Lambda) = H^{(0)}(\Lambda) - \sum_{l=1}^{\infty} J(l) \sum_{\substack{n+1,\ldots,n+l\\ \in\Lambda}} P^{(0)}(n+1)\ldots P^{(0)}(n+l), \quad (58)$$

where $H^{(0)}(\Lambda)$ is an operator in $\mathcal{H}^{(0)}(\Lambda)$, representing the short-range forces between particles in $\mathcal{H}_0^{(0)}$-class states, and $J(l)$ is a positive parameter, representing the strength of the l-body interaction between a chain of consecutive atoms in those states. We also assume that the parameters $J(l)$ possess the following properties which ensure that (a) the conditions (20) and (21), concerning the uniform boundedness of the energy density and the 'surface' character of the interaction between bounded regions are fulfilled; and (b) the energy of interaction between the particles inside and outside a bounded region is stronger than a surface effect:

$$\sum_1^{\infty} J(l) \quad \text{is finite} \quad (59)$$

and

$$\sum_1^{N} lJ(l) = 0(N^{1-\alpha}) \quad \text{for some } \alpha \text{ between 0 and 1.} \quad (60)$$

These conditions are fulfilled if, for example, $J(l) = \text{const} \times l^{-(1+\alpha)}$. In general, (60) implies that

$$\sum_1^{\infty} lJ(l) \quad \text{is infinite.} \quad (61)$$

In order to construct states of Σ that are locally, but not globally, stable, we introduce the auxiliary system, $\Sigma^{(0)}$, consisting of an assembly of atoms on X with atomic state space $\mathcal{H}_0^{(0)}$ and local Hamiltonian $H^{(0)}(\Lambda)$. The states of $\Sigma^{(0)}$ may therefore be identified with the subset, $\Omega^{(0)}$, of states of Σ whose restrictions to the bounded regions, Λ, are given by density matrices in $\mathcal{H}^{(0)}(\Lambda)$. Furthermore, as the projectors $P^{(0)}(n)$ reduce to unity on the

$\mathscr{H}_0^{(0)}$-class atomic states, it follows from (58) that $H(\Lambda)$ reduces to $H^{(0)}(\Lambda)$, plus a constant, when Σ is constrained to the states $\Omega^{(0)}$. Consequently, the GTS states of $\Sigma^{(0)}$ are simply the minima of the free energy density $\hat{f}$ of Σ over the reduced state space $\Omega^{(0)}$. We shall now show that these states of Σ are locally stable at all temperatures, but not globally stable at sufficiently high ones.

Thus, we consider the stability of a state $\rho^{(0)}$ of Σ that is GTS for $\Sigma^{(0)}$ or, equivalently, a minimum of the free energy density, $\hat{f}$, of Σ over the reduced state space $\Omega^{(0)}$. Since $\rho^{(0)}$ is GTS for $\Sigma^{(0)}$ and since the interactions in that system, as represented by $\{H^{(0)}(\Lambda)\}$, are of short range, it follows that $\rho^{(0)}$ is also LTS for $\Sigma^{(0)}$. Thus, if ρ' is another state of $\Sigma^{(0)}$ obtained by local modification of $\rho^{(0)}$, then

$$\Delta\hat{F}^{(0)}(\rho^{(0)} \mid \rho') \geqslant 0,$$

where $\Delta\hat{F}^{(0)}$ is the incremental free energy for $\Sigma^{(0)}$. Equivalently, $\Delta\hat{F}^{(0)}$ is the incremental free energy, $\Delta\hat{F}$, of Σ over the reduced state space $\Omega^{(0)}$, since $H(\Lambda)$ and $H^{(0)}(\Lambda)$ differ by a constant over those states. Hence, the last inequality signifies that

$$\Delta\hat{F}(\rho^{(0)} \mid \rho') \geqslant 0 \quad \text{for } \rho' \text{ in } \Omega^{(0)}.$$

Hence, to prove that $\rho^{(0)}$ is LTS for Σ, it remains for us to show that this inequality is valid even when ρ' does not lie in $\Omega^{(0)}$. For this, we start by considering the modification of $\rho^{(0)}$ due to the transfer of a single atom, from an $\mathscr{H}_0^{(0)}$- to an $\mathscr{H}_0^{(1)}$-class state. Since the number of chains of l consecutive atoms that include any one fixed atom is l, it follows that the resultant increment in energy, due to many-particle interactions, is $\sum_1^\infty lJ(l)$, which is infinite, by (61). This argument may be extended† to show that, if ρ' is any state obtained by local modification of $\rho^{(0)}$, then the incremental energy $\Delta\hat{E}(\rho^{(0)} \mid \rho')$ is infinite. On the other hand, the entropy increment $\Delta\hat{S}(\rho^{(0)} \mid \rho')$ is finite, since as noted just before (41), this is true for local modifications of any state. Consequently, $\Delta\hat{F}(\rho^{(0)} \mid \rho')$ is positive (and infinite!), and therefore $\rho^{(0)}$ is LTS at any temperature.

To prove that $\rho^{(0)}$ is not a GTS state of Σ at sufficiently high temperatures, we note that it follows from the condition (20) on the local energy density that the functional $\hat{e}(\rho)$ is uniformly bounded, i.e.

$$|\hat{e}(\rho)| < C \quad \text{for all states } \rho, \tag{62}$$

where C is a finite constant. This evidently implies that the entropy must play a dominant role in determining the GTS states at high temperatures. Suppose now that the dimensionalities of the atomic space $\mathscr{H}_0$ and its subspace $\mathscr{H}_0^{(0)}$ are m and m_0 ($<m$) respectively. Then, if $\bar{\rho}$ is the state of Σ in

† See Sewell (1977*b*) for details.

which the atoms are uncorrelated with each one in its state of maximum entropy, i.e. $\bar{\rho} = \bigotimes_n \bar{\rho}_n$, where $\bar{\rho}_n$ corresponds to the density matrix $m^{-1}I_n$, I_n being the unit operator in $\mathscr{H}_n$, then

$$\hat{s}(\bar{\rho}) = k \ln m. \tag{63}$$

It also follows easily from the subadditivity of entropy and the definition (30) of $\hat{s}$ that $\hat{s}(\bar{\rho})$ is the maximum value of this functional. Likewise, the maximum value of $\hat{s}$ over the $\Omega^{(0)}$-class states is $k \ln m_0$. On combining this observation with the formulae (62) and (63), we see that

$$\hat{f}(\rho^{(0)}) \geq -C - kT \ln m_0$$

and

$$\hat{f}(\bar{\rho}) \leq C - kT \ln m.$$

Hence $\hat{f}(\bar{\rho}) < \hat{f}(\rho^{(0)})$ if $kT > 2C/\ln(m/m_0)$, which proves that $\rho^{(0)}$ is not GTS at sufficiently high temperatures.

One may summarize this result by saying that, whereas the many-body forces guarantee the stability of $\rho^{(0)}$ against all local modifications, the deficiency in entropy of this state renders it unstable against certain global changes at sufficiently high temperatures.

Note. We remark that the dynamics of Σ in its $\Omega^{(0)}$-class states is governed by the interactions represented by the local Hamiltonians $\{H^{(0)}(\Lambda)\}$, since the projectors $P^{(0)}(\Lambda)$ reduct to unity in those states.

3.7. The KMS fluctuation-dissipation condition

We recall from §3.1 that the canonical state of a finite system at temperature T is completely characterized by the KMS condition, i.e.

$$\rho(A_t B) = \rho(B A_{t+i\hbar\beta}), \qquad \beta = (kT)^{-1}, \tag{13}$$

for all observables A and B. As first appreciated by Haag, Hugenholtz and Winnink (1967), this condition, as applied to the local observables of an infinite system, represents a key property of its equilibrium states too. Before examining its physical significance, we first remark that eqn (13) is merely formal for infinite systems, since complex-time translates of observables are not always definable. However, one can cast it into a precise mathematical form, first noting that (cf. §2.4.5), for real t, $\rho(A_t B)$ and $\rho(BA_t)$ are defined in terms of the GNS representation $\hat{A}_t$, $\hat{B}$ of A_t, B and the cyclic vector Ψ for the state ρ as $(\Psi, \hat{A}_t \hat{B} \Psi)$ and $(\Psi, \hat{B}\hat{A}_t \Psi)$, respectively. The KMS condition may then be precisely stated in the following form.

(KMS) *For each pair of local observables A, B, there is a function $F(z)$ of the complex variable z, that is analytic in the strip* $\text{Im}\, z \in (0, \hbar\beta)$ *and continuous on its boundaries, such that*

$$F(t) = \rho(AB_t) \quad \text{and} \quad F(t + i\hbar\beta) = \rho(A_t B) \quad \text{for real } t.$$

Turning now to the physical significance of this condition, we observe that the following results, which will be discussed below, have been proved.

(1)† *The KMS condition is equivalent to that of local thermodynamical stability.*

(2)‡ *This condition characterizes those states for which the system behaves as a thermal reservoir, in the sense that it drives any finite system, S, to which it is weakly and locally coupled, into thermal equilibrium at temperature Σ.*

(3)§ *The coupling of Σ to the finite system S induces transitions between the eigenstates of the latter system at rates that satisfy the principle of detailed balance. Thus if P_j is the canonical occupation probability of the eigenstate ϕ_j of S at temperature T, and W_{jk} is the rate of transitions from ϕ_j to another eigenstate ϕ_k, due to the Σ-S coupling, then*

$$W_{jk}P_j = W_{kj}P_k, \tag{64}$$

i.e.

$$W_{jk} = W_{kj} \exp \beta(E_j - E_k), \tag{64$'$}$$

where E_j and E_k are the energy levels of ϕ_j and ϕ_k, respectively.

The property (1) is a generalization, at the local level, of the equivalence between the KMS and local stability conditions for finite systems, discussed in §3.1 and proved in Appendix A. That it is at the local level that this generalization is operative is quite natural, since the KMS condition concerns correlations between local observables.

The properties (2) and (3) signify that the KMS condition specifies those states for which the infinite system Σ behaves as a thermal reservoir, in the sense of the *Zeroth Law of Thermodynamics*.

We shall now sketch the proof of (2) and (3) for the simple case where S is a two-level atom: the general proof follows from an analogous argument (cf. Kossakowski *et al*. (1977)).

3.7.1. *The reservoir property and detailed balance: a simple example*

We consider the situation where the infinite system Σ is coupled to a two-level atom, S. We assume that the eigenstates of S are given by the normalized vectors ϕ_1 and ϕ_2 in a two-dimensional space $\mathcal{K}$, and that the corresponding energy levels are E_1 and E_2, respectively. We assume that the interaction energy between Σ and S is of the form

$$H_I = \lambda(C^* \otimes c + C \otimes c^*), \tag{65}$$

where λ is a dimensionless coupling constant, C is an element of the algebra

† See Araki and Sewell (1977), and Sewell (1977*a*, 1980).
‡ See Kossakowski *et al*. (1977).
§ See Kossakowski *et al*. (1977).

of local observables of Σ, and c^*, c are the raising and lowering operators for S defined by

$$c^*\phi_1 = \phi_2; \qquad c^*\phi_2 = 0; \qquad c\phi_1 = 0; \qquad c\phi_2 = \phi_1. \tag{66}$$

We also assume that Σ and S are initially prepared, independently of one another, in states ρ and σ, respectively, and then coupled together via the interaction H_I. Thus, the composite system $(\Sigma + S)$ evolves from an initial state $\rho \otimes \sigma$ under the effects of this interaction. We assume that ρ is a stationary state for the system Σ, when that is isolated, and that this state satisfies the condition

$$\rho(C) = 0. \tag{67}$$

In fact, one can always arrange that this last condition is fulfilled by adding a suitable constant to C in (65) and compensating this by a counterterm in the Hamiltonian for S.

We consider the dynamics of S induced by the evolution of the composite system $(\Sigma + S)$. This may be formulated as follows. An observable A of S corresponds to the observable $I \otimes A$ of the composite system. Consequently, if $(\rho \otimes \sigma)_t$ is the time-evolute of $(\Sigma + S)$ from its initial state $(\rho \otimes \sigma)$, the time-dependent expectation value of the observable, A, of S is $(\rho \otimes \sigma)_t\,(I \otimes A)$. This means that the time-dependent state σ_t of S is given by

$$\sigma_t(A) = (\rho \otimes \sigma)_t(I \otimes A). \tag{68}$$

Thus, σ_t depends both on the full dynamics of the composite system and on the initial state, ρ, of Σ.

We consider the evolution of S in the case where the system is weakly coupled to Σ. In this case, the transition rates between the eigenstates of S are proportional, in Born approximation, to the square of the coupling constant, λ, governing the Σ–S interaction. This suggests that the 'natural' time-scale for the evolution of S is essentially λ^{-2}. Accordingly we formulate this evolution in terms of the rescaled time variable† $\tau = \lambda^2 t$, defining the time-dependent state of S on the τ-scale as

$$\bar{\sigma}_\tau^{(\lambda)} \equiv \sigma_{\lambda^2 t}, \quad \text{with} \quad \tau \equiv \lambda^2 t. \tag{69}$$

Now the evolution of this state becomes very simple, on the τ-scale, in the weak-coupling limit where $\lambda \to 0$. For it has been proved‡ that, in this limit, the occupation probabilities, P_1 and P_2, of the eigenstates, ϕ_1 and ϕ_2, of S evolve according to the *Pauli master equation*, subject to certain viable conditions on the decay of the time-correlation functions of Σ for the state ρ.

† The idea of rescaling time in this way is due originally to Van Hove (1955).

‡ The proof is covered by Davies's (1974) treatment of master equations, which substantiates ideas by Van Hove (1955). We remark here that the Van Hove limit ($\lambda \to 0, \lambda^2 t \equiv \tau$ fixed) is needed in order to eliminate 'memory effects', which would lead to an integro-differential equation for P_1 and P_2, instead of (70).

This master equation takes the form

$$\frac{dP_1}{d\tau} = -\frac{dP_2}{d\tau} = W_{21}P_2 - W_{12}P_1, \tag{70}$$

where the transition rates W_{21} and W_{12} are expressed in terms of time-correlation functions for the isolated system Σ by the Born approximation, corresponding to the form of the interaction given by (65), i.e.

$$\left.\begin{aligned} W_{21} &= \hbar^{-2}\int_{-\infty}^{\infty} dt\rho(C_t^*C)e^{-i\omega t} \\ \text{and} \quad W_{12} &= \hbar^{-2}\int_{-\infty}^{\infty} dt\rho(CC_t^*)e^{-i\omega t}, \end{aligned}\right\} \tag{71}$$

where

$$\hbar\omega = E_1 - E_2. \tag{72}$$

It now follows from the master equation (70) that P_1 and P_2 tend to terminal values, $\bar{P}_1$ and $\bar{P}_2$, as $\tau \to \infty$, where

$$\bar{P}_1 = W_{21}/(W_{21} + W_{12}), \quad \text{and} \quad \bar{P}_2 = W_{12}/(W_{21} + W_{12}).$$

These terminal probabilities are therefore the canonical ones, for temperature T, if and only if

$$W_{21}e^{-\beta E_2} = W_{12}e^{-\beta E_1}, \tag{73}$$

i.e. by (72), if

$$W_{21} = W_{12}e^{-\beta\hbar\omega}. \tag{73$'$}$$

This is the *detailed balance condition* (cf. (64)′). Further, in view of (71), this condition is equivalent to

$$\int_{-\infty}^{\infty} dt\rho(C_t^*C)e^{-i\omega t} = \int_{-\infty}^{\infty} dt\rho(CC_t^*)e^{-i\omega t - \beta\hbar\omega}, \tag{74}$$

which is the Fourier transform, for frequency ω, of the KMS relation

$$\rho(C_t^*C) = (CC_{t+i\hbar\beta}^*). \tag{75}$$

We see, then, that the terminal probabilities, $\bar{P}_1$ and $\bar{P}_2$, are the canonical ones if and only if the state ρ satisfies the Fourier transform, for frequency ω, of the KMS relation (75). The condition for these terminal probabilities to be canonical for *any* two-level atom coupled to Σ is therefore that the KMS relation (75) is valid for all elements C of its algebra of local observables that satisfy (67). This is, in fact, equivalent to the *full KMS condition*,

$$\rho(A_tB) = \rho(BA_{t+i\hbar\beta}), \tag{13}$$

since, on the one hand, (75) is left unchanged by the addition of an arbitrary constant to C, while, on the other hand, one may easily pass† from (75) to (13) by virtue of the linearity of $\rho(A_tB)$ and $\rho(BA_t)$ in A and B.

Thus, the KMS relation (13) is the condition for the terminal occupation probabilities, $\bar{P}_1$ and $\bar{P}_2$, of the eigenstates of any two-level atom, weakly and locally coupled to Σ, to be canonical. It is therefore the condition for the terminal value of the *diagonal part* of the density for S to be canonical. This result may, in fact, be extended‡ to the full density matrix for S, via a master equation involving its off-diagonal elements as well as its diagonal ones.

We therefore conclude that the KMS relation represents the condition on the state of Σ for which this system behaves as a thermal reservoir, with respect to two-level atoms, in that it drives these into canonical equilibrium states at temperature T, when locally and weakly coupled to them. Furthermore, it is the condition that the dynamics of such atoms, induced by their coupling to Σ, satisfies the principle of detailed balance.

Comment. As we have just seen, a canonical (i.e. KMS) state of a finite system is stable in the face of coupling to a thermal reservoir, as represented by an infinite system in a KMS state at the same temperature. The essential reason for this is that the KMS property of the reservoir induces a dynamics in the finite system that conforms to the principle of detailed balance and so ensures the (dynamical) stability of its canonical state. The same argument may also be employed to establish that a locally stable (i.e. KMS) state of an *infinite* system Σ is dynamically stable, i.e. infinitely long-lived, in the face of couplings to a reservoir, Σ_R, at the same temperature. Here, however, the reservoir should be taken to be 'infinitely large' by comparison with Σ and its coupling to different regions of this system should be uncorrelated. This state of affairs may be represented by the assumption that Σ_R consists of an infinite set of independent reservoirs $\{\Sigma_R^{(n)}\}$, and that these are coupled to different parts of Σ occupying disjoint finite spatial regions.§ Under appropriate conditions on the dynamics of these reservoirs, the resultant evolution of Σ is then given by a master equation satisfying the detailed balance condition, and its KMS states are then dynamically stable.

3.8. Conclusions concerning equilibrium states

The results concerning the thermodynamical stability and KMS conditions, discussed in the previous Sections, provide us with the basis for an operational definition of equilibrium states of infinite systems. In fact, we base our definition of these states on the assumption that they are

† For, given any local observables A and B, $\rho(A_tB)$ and $\rho(BA_t)$ may be expressed in the forms $\sum_{j=1}^4 l_j\rho(C_{j,t}^*C_j)$ and $\sum_{j=1}^4 l_j\rho(C_jC_{j,t}^*)$, respectively, where C_1, $C_2 = A \pm B$, C_3, $C_4 = A \mp iB$ and $l_1 = -l_2 = -il_3 = il_4 = \frac{1}{4}$.

‡ See Kossakowski *et al*. (1977).

§ Cf. Hepp and Lieb (1973); Martin (1977).

characterized by the requirements of the Second and Zeroth Laws of Thermodynamics, in the sense that (a) they are thermodynamically stable, at both global and local levels; and (b) they possess the reservoir property, described in §3.7. Now as we have seen in the previous Sections, this latter property reduces to that of KMS or, equivalently, of local stability; and this is covered by the global stability requirement of (a). Hence, the characterization of equilibrium by the conditions (a) and (b) implies that *the equilibrium states are the globally stable ones*. Furthermore, we draw the important conclusion from this definition and the results of §3.5 that *the equilibrium states of an infinite system can exhibit the phenomena of macroscopic degeneracy and symmetry breakdown*, which are essential for a theory of the phase structure of matter.

As regards the locally, but not globally, stable states, supported by certain systems with long-range forces, we remark that these satisfy the KMS condition and therefore possess the thermal reservoir property required by the Zeroth Law. They therefore simulate true equilibrium states in two important respects, namely that they enjoy the reservoir property and that they are stable with respect to local modifications. This strongly suggests that these are *metastable states*. In fact, this will be our conclusion in Chapter 6, when we examine the matter in more detail.

Continuous systems. The picture of equilibrium, and of metastability, presented here has been obtained from a treatment of lattice systems. The corresponding theory of continuous systems is technically more difficult, because of problems connected with unbounded local observables, and has not yet been worked out in full detail. Nevertheless, the rigorous results available for these systems lead essentially to the same picture as the one we have described for lattice systems.

Let us conclude this chapter by reviewing the results that have been established for thermodynamical stability and KMS conditions for continuous systems.

First we note that the whole of the theory of *global stability* has been extended to a rather wide class of continuous systems.† Essentially, these are the systems whose interactions are sufficiently repulsive at short distances to prevent the collapse of an infinity of particles into a finite volume, and fall off fast enough at large distances to ensure that energy is an extensive variable. Thus, the theory is applicable to continuous systems of particles with hard cores and short-range forces, and even to systems with interatomic forces of the Lenard-Jones type.‡

As regards *local stability*, it is a straightforward matter to extend its

† See Robinson (1971).
‡ See A. Pflug (Thesis, University of Vienna, 1978).

formulation† to continuous systems of particles with hard cores and suitably regular interactions. Here again, the translationally invariant locally stable states are globally stable, while systems with suitably long-range forces can support states that are locally, but not globally, stable.

The theory of the *KMS condition* and its characterization of the reservoir properties is applicable to continuous systems, too, subject to the general assumptions concerning their dynamics, described in §2.4.5. As regards the relationship between this condition and global stability for infinite systems, there is little in the way of general results to be found in the literature. However, it is not difficult to check that the method employed by Lanford and Robinson (1968*b*) to prove that globally stable states of lattice systems satisfy the KMS condition may be extended‡ to establish the same result for continuous systems of particles with hard cores, as formulated by Miracle-Sole and Robinson (1970).

Thus, the theory of thermodynamical stability and the KMS property extends to at least a wide class of continuous systems. The operational definition of equilibrium states as the globally stable ones is therefore applicable to these systems, too; and, as in the case of lattice systems, this admits the phenomena of macroscopic degeneracy and symmetry breakdown, required for a theory of phase structure. As for the states that are locally, but not globally, stable, these may again be interpreted as metastable.

Note on microcanonical description. The formulation of equilibrium states employed here has been the canonical one. However, one can also formulate these states in microcanonical terms, characterizing them by the condition that they maximize the entropy density $\hat{s}$, given the value of the energy density $\hat{e}$. For, if ρ is an equilibrium state and ρ' is any other translationally invariant state with the same energy density, then it follows from the global stability of ρ that

$$\hat{e}(\rho) - T\hat{s}(\rho) \leqslant \hat{e}(\rho') - T\hat{s}(\rho').$$

Therefore, as $\hat{e}(\rho) = \hat{e}(\rho')$, it follows that $\hat{s}(\rho) \geqslant \hat{s}(\rho')$, which means that equilibrium states maximize the entropy density, given the value of the entropy density.

To prove the converse, we define $s(e)$ as the maximum value§ of $\hat{s}$, given

† Here, as in the formulation of the global free energy functional, one must define the local kinetic energy observable $T(\Lambda)$ with Neumann, rather than Dirichlet, boundary conditions in order that it possesses the additivity property that $\rho(T(\Lambda_1 \cup \Lambda_2)) = \rho(T(\Lambda_1)) + \rho(T(\Lambda_2))$. For a discussion of these boundary conditions, see Robinson (1971).

‡ One can also prove the same thing for a significant class of continuous systems by a method due to Narnhofer and Sewell (1980; Appendix 2) which, though devised for the specific case of free fermions, may in fact be seen to be applicable in a much wider context.

§ The attainment of a maximum value by $\hat{s}$, at given energy density, is guaranteed by its upper semicontinuity.

that the energy density is e. This is a concave function of e since it follows from the definition of $s(e)$, together with the affine properties of $\hat{e}$ and $\hat{s}$ that if ρ_1, ρ_2 are states that maximize $\hat{s}$ for energy densities e_1, e_2, respectively, and $0<\lambda<1$, then

$$\lambda e_1 + (1-\lambda)e_2 = \hat{e}(\lambda\rho_1 + (1-\lambda)\rho_2)$$

and

$$\begin{aligned} s(\lambda e_1 + (1-\lambda)e_2) &\geqslant \hat{s}(\lambda\rho_1 + (1-\lambda)\rho_2) \\ &= \lambda\hat{s}(\rho_1) + (1-\lambda)\hat{s}(\rho_2) = \lambda s(e_1) + (1-\lambda)s(e_2). \end{aligned}$$

The concavity of $s(e)$ implies that $\mathrm{d}s(e)/\mathrm{d}e$ is a non-increasing function of e, and therefore that

$$s(e') \leqslant s(e) + \frac{\mathrm{d}s(e)}{\mathrm{d}e}(e'-e)$$

for arbitrary energy densities e and e'. Consequently, defining the temperature T, corresponding to e, as $(\mathrm{d}s(e)/\mathrm{d}e)^{-1}$, in the standard way, it follows that

$$e - Ts(e) \leqslant e' - Ts(e').$$

Thus, if ρ maximizes $\hat{s}$ for energy density e and if ρ' is another state of energy density e', then $e=\hat{e}(\rho)$, $s(e)=\hat{s}(\rho)$, $e'=\hat{e}(\rho')$ and $s(e')\geqslant\hat{s}(\rho')$, so that the above inequality for $e-Ts(e)$ implies that

$$\hat{e}(\rho) - T\hat{s}(\rho) \leqslant \hat{e}(\rho') - T\hat{s}(\rho'),$$

i.e. that ρ is globally stable and therefore an equilibrium state.

Appendix A
Thermodynamical stability and the KMS condition for finite systems

We shall devote this Appendix to proving that the canonical state ρ_c of a finite system is equivalently characterized by the thermodynamical stability condition (5) and by the KMS condition (13), as stated in §3.1. For simplicity, we shall confine the proof to the case where the Hilbert space $\mathscr{H}$ is finite-dimensional.

Proof of the thermodynamical stability formula (5).
By eqns (1)–(4),

$$\hat{F}(\rho) - \Phi(T) = kT\,\mathrm{Tr}(\rho\ln\rho - \rho\ln\rho_c),$$

from which it follows that $\hat{F}(\rho_c)=\Phi(T)$, and that, in order to establish the relation (5), it suffices to prove that

$$\mathrm{Tr}(\rho\ln\rho - \rho\ln\rho_c) > 0 \quad \text{for } \rho\neq\rho_c. \tag{A.1}$$

Let $\{\psi_n\}$ and $\{p_n\}$ be the eigenvectors and corresponding eigenvalues of ρ_c. Then p_n is the occupation probability of ψ_n for this state, and the LHS of (A.1) is equal to

$$\sum_n [(\psi_n, (\rho \ln \rho)\psi_n) - (\psi_n, \rho\psi_n)\ln p_n]. \tag{A.2}$$

We shall prove the positivity of this expression, for $\rho \neq \rho_c$, in two steps, both based on the fact that $x \ln x$ is a strictly convex function of the positive real variable x.

Step 1. We prove that, for any normalized vector ψ,

$$(\psi, (\rho \ln \rho)\psi) \geqslant (\psi, \rho\psi)\ln(\psi, \rho\psi), \tag{A.3}$$

equality occurring if and only if ψ is an eigenvector of ρ. The proof runs as follows. Let $\{w_n\}$ be the eigenvalues of ρ, and let P_n be the projection operator for the subspace of $\mathcal{H}$ spanned by the eigenvectors corresponding to w_n. Thus,

$$\rho = \sum w_n P_n$$

and

$$\rho \ln \rho = \sum (w_n \ln w_n) P_n.$$

Consequently,

$$(\psi, \rho\psi) = \sum c_n w_n, \tag{A.4}$$

and

$$(\psi, (\rho \ln \rho)\psi) = \sum c_n w_n \ln w_n, \tag{A.5}$$

where

$$c_n = (\psi, P_n\psi), \tag{A.6}$$

the probability that ρ takes the value w_n for the vector state ψ. Since $x \ln x$ is a strictly convex function† of the positive variable x,

$$\sum_n c_n w_n \ln w_n \geqslant \left(\sum_n c_n w_n\right) \ln\left(\sum_n c_n w_n\right),$$

equality occurring if and only if one of the c_n's is unity and the rest are zero. By eqns (A.4) and (A.5), this result is equivalent to the required one, (A.3).

Step 2. Now let $\{\psi_n\}$ and $\{\bar{p}_n\}$ be the eigenvectors and corresponding eigenvalues of ρ_c, and define

$$p_n = (\psi_n, \rho\psi_n), \tag{A.7}$$

the occupation probability of ψ_n for the state ρ. Then it follows from (A.3)

† The strict convexity of $x \ln x$ follows from the fact that its second derivative, x^{-1}, is positive.

that the expression (A.2) is greater than or equal to

$$\sum_n p_n \ln p_n/\bar{p}_n, \tag{A.8}$$

equality occurring if and only if $\{\psi_n\}$ are eigenvectors of ρ. Hence, in order to prove the positivity of (A.2) for $\rho \neq \rho_c$, it suffices to establish that of (A.8) unless $p_n = \bar{p}_n$ for all n. For this purpose, we note that (A.8) is identical to

$$\sum_n \bar{p}_n((p_n/\bar{p}_n)\ln(p_n/\bar{p}_n)),$$

and, if $p_n \neq \bar{p}_n$ for all n, it follows from the strict convexity of $x \ln x$ that this expression exceeds

$$\left(\sum_n \bar{p}_n(p_n/\bar{p}_n)\right)\ln\left(\sum_n \bar{p}_n(p_n/\bar{p}_n)\right) \equiv \left(\sum_n p_n\right)\ln\left(\sum_n p_n\right),$$

which is zero, since $\sum_n p_n = 1$. Hence (A.8) is positive unless $p_n \equiv \bar{p}_n$, which is the required result.

Proof that the KMS condition characterizes ρ_c

We first note that the KMS condition (13) is equivalent to

$$\rho(AB) = \rho(BA_{i\hbar\beta}), \quad \text{for all observables } A, B. \tag{A.9}$$

For on the one hand, (13) trivially implies (A.9); while on the other hand, the latter condition yields (13) when A is replaced by A_t. Thus, we may express the KMS condition in the form

$$\mathrm{Tr}(\rho AB) = \mathrm{Tr}(\rho BA_{i\hbar\beta}) \equiv \mathrm{Tr}(\rho B\mathrm{e}^{-\beta H}A\mathrm{e}^{\beta H}).$$

In view of the cyclical property of the Trace($\mathrm{Tr}(A_1A_2A_3) \equiv \mathrm{Tr}(A_2A_3A_1)$), this is equivalent to

$$\mathrm{Tr}((B\rho - \mathrm{e}^{\beta H}\rho B\mathrm{e}^{-\beta H})A) = 0.$$

This condition, as applied to all operators A, B in $\mathscr{H}$, is valid if and only if

$$\mathrm{e}^{\beta H}\rho B\mathrm{e}^{-\beta H} = B\rho,$$

i.e.

$$\mathrm{e}^{\beta H}\rho B = B\rho\mathrm{e}^{\beta H}, \quad \text{for all operators } B \text{ in } \mathscr{H}. \tag{A.10}$$

We see immediately from the definition (1) of the canonical state that (A.10) is satisfied by $\rho = \rho_c$. Conversely, if ρ satisfies (A.10), then it follows on putting $B = I$ that ρ commutes with $\mathrm{e}^{\beta H}$. The condition (A.10) therefore implies that

$$\mathrm{e}^{\beta H}\rho B = B\mathrm{e}^{\beta H}\rho \quad \text{for all operators } B \text{ in } \mathscr{H}.$$

Hence, $\mathrm{e}^{\beta H}\rho$ commutes with all operators in $\mathscr{H}$ and is therefore of the form cI, where c is a scalar. This means that $\rho = c\mathrm{e}^{-\beta H}$ and, as $\mathrm{Tr}\,\rho = 1$, ρ must be the canonical density matrix $\rho_c \equiv e^{-\beta H}/\mathrm{Tr}\, e^{-\beta H}$. In other words, ρ_c is the only density matrix that satisfies the KMS condition.

Appendix B
The compactness property (*P*)

In order to prove that the Bolzano–Weierstrass theorem may be extended so as to yield the compactness property (P) of the states of Σ, we first establish that the algebra of local observables, $\mathcal{A}$, of this system consists of linear combinations of a certain sequence $\{A_n\}$ of its elements. We prove this property of $\mathcal{A}$ by noting that, since the atomic Hilbert space $\mathcal{H}_0$ is finite dimensional, the operators in that space are given by linear combinations of a finite basis set $(B^{(1)}, \ldots, B^{(N)})$ say. Correspondingly, the observables at a lattice site x are linear combinations of a finite basis set $(B_x^{(1)}, \ldots, B_x^{(N)})$. Thus, it follows from the definition of the algebra $\mathcal{A}$ that this consists of finite linear combinations of terms $B_{x_1}^{(l_1)} \ldots B_{x_k}^{(l_k)}$, where $l_1, \ldots, l_k$ run through the values $(1, \ldots, N)$, and the x's through the positions of the lattice sites. These products therefore form a denumable basis $\{A_n\}$ for $\mathcal{A}$.

Suppose now that $S \equiv \{\rho_n\}$ is a sequence of states of Σ. Then it follows from the Bolzano–Weierstrass theorem that S has a subsequence, S_1, which converges on A_1; that S_1 has a subsequence S_2, which converges on A_2, etc. The sequence S therefore has subsequences $S_1, S_2, \ldots, S_n, \ldots$ such that $S_1 \supset S_2 \supset \ldots \supset S_n, \ldots$, and S_n converges on $A_1, \ldots, A_n$. Thus, denoting the sequence S_n by $(\rho_{n1}, \rho_{n2}, \ldots)$, we see that S_n contains all but a finite number of the elements of the subsequence† $\bar{S} \equiv (\rho_{11}, \rho_{22}, \ldots, \rho_{nn}, \ldots)$ of S. Therefore $\bar{S}$ converges on each of the basis elements $\{A_n\}$ of the algebra $\mathcal{A}$ and hence on $\mathcal{A}$ itself. This proves that any sequence, S, of states has a convergent subsequence, $\bar{S}$, and therefore establishes the compactness property (P).

Appendix C
Infinite volume limits

We shall devote this Appendix to the proof of

(a) the convergence of the canonical free energy density, $\Phi_\Lambda(T)/|\Lambda|$, and of the energy and entropy density functionals, $\hat{E}_\Lambda(\rho)/|\Lambda|$ and $\hat{S}_\Lambda(\rho)/|\Lambda|$, as Λ increases to infinity;

(b) the affine property of the resultant limit functionals, $\hat{e}(\rho)$ and $\hat{s}(\rho)$; and

(c) the continuity of $\hat{e}(\rho)$ and the upper semicontinuity of $\hat{s}(\rho)$.

† The construction of the subsequence $\bar{S}$ in this way is usually called the 'diagonalization procedure', and is standard in Analysis (cf. Dieudonné 1960, §9.13).

For simplicity,† we shall confine our proof of (a) to situations where Λ increases over a sequence $\{\Lambda_n\}$ of cubes, such that Λ_{n+1} is the union of 2^d copies $\{\Lambda_{n,j} \mid j=1,\ldots,n\}$ of Λ_n, d being the dimensionality of the lattice X. In this case, it follows from the surface condition (21) that the interaction between the cubes $\{\Lambda_{n,j}\}$ possesses the key property that

$$\left\| H(\Lambda_{n+1}) - \sum_{j=1}^{2^d} H(\Lambda_{n,j}) \right\| \Big/ |\Lambda_{n+1}|^{1-\varepsilon} \to 0 \quad \text{as} \quad n \to \infty.$$

Since $|\Lambda_{n+1}|$ is proportional to $2^{(n+1)d}$, this implies that, for some finite constant c,

$$\left\| H(\Lambda_{n+1}) - \sum_{j=1}^{2^d} H(\Lambda_{n,j}) \right\| \Big/ |\Lambda_{n+1}| < c/2^{nd\varepsilon} \tag{C.1}$$

Convergence of $\Phi_\Lambda(T)/|\Lambda|$. By the thermodynamical stability condition (5) for finite systems,

$$\Phi_\Lambda(T) \leqslant \mathrm{Tr}_\Lambda(\rho_\Lambda H(\Lambda) + kT\rho_\Lambda \ln \rho_\Lambda)$$

for any density matrix ρ_Λ in $\mathscr{H}(\Lambda)$. On applying this inequality to the case where $\Lambda = \Lambda_{n+1}$ and ρ_Λ is the tensor product $\bigotimes_{j=1}^{2^d} \bar{\rho}_{\Lambda_{n,j}}$ of the canonical density matrices $\bar{\rho}_{n,j}$ ($\equiv \mathrm{e}^{-\beta H(\Lambda_{n,j})}/\mathrm{Tr}(\text{idem})$) for the cubes $\Lambda_{n,j}$ we see that

$$\Phi_{\Lambda_{n+1}}(T) \leqslant 2^d \Phi_{\Lambda_n}(T) + \mathrm{Tr}\left(\rho_{\Lambda_{n+1}}\left(H(\Lambda_{n+1}) - \sum_1^{2^d} H(\Lambda_{n,j})\right)\right).$$

By dividing this inequality by $|\Lambda_{n+1}|$ and using (C.1), we see that

$$\Phi_{\Lambda_{n+1}}(T)/|\Lambda_{n+1}| \leqslant \Phi_{\Lambda_n}(T)/|\Lambda_n| + c/2^{nd},$$

or equivalently,

$$\Phi_{\Lambda_n}(T)/|\Lambda_n| - c\sum_{l=1}^{n-1} 2^{-ld\varepsilon} \text{ decreases monotonically with } n. \tag{C.2}$$

Now the sum in this quantity converges as $n \to \infty$. Hence, in order to establish the convergence of $\Phi_{\Lambda_n}(T)/|\Lambda_n|$ it suffices to prove that it has a lower bound, which does not depend on n. To this end, we note that it follows from the definition (34) of $\Phi_\Lambda(T)$ that

$$|\Phi_\Lambda(T)| \leqslant \|H(\Lambda\| + kT \ln \mathrm{D}(\Lambda),$$

where $\mathrm{D}(\Lambda)$ is the dimensionality of $\mathscr{H}(\Lambda)$. Consequently, as $\mathrm{D}(\Lambda) = v^{|\Lambda|}$, where v is the dimensionality of the atomic space $\mathscr{H}_0$, and as $\|H(\Lambda)\| < C|\Lambda|$ for some finite C, by (20), it follows that

$$|\Phi_\Lambda(T)|/|\Lambda| \leqslant C + kT \ln v,$$

which proves that the sequence $\{\Phi_{\Lambda_n}(T)/|\Lambda|\}$ is lower-bounded and therefore, in view of (C.2), convergent.

† For the generalization to a much larger class of increasing bounded regions, see Ruelle (1969a).

Convergence of $\hat{E}_\Lambda(\rho)/|\Lambda|$. It follows from the definition (24) of $\hat{E}_\Lambda(\rho)$, that

$$\left|\hat{E}_{\Lambda_{n+1}}(\rho) - \sum_{j=1}^{2^d} \hat{E}_{\Lambda_{n,j}}(\rho)\right| \leqslant \left\| H(\Lambda_{n+1}) - \sum_{j=1}^{2^d} H(\Lambda_{n,j})\right\|.$$

For translationally invariant states $\rho, E_{\Lambda_{n,j}}(\rho) \equiv E_{\Lambda_n}(\rho)$, and consequently

$$|\hat{E}_{\Lambda_{n+1}}(\rho) - 2^d \hat{E}_{\Lambda_n}(\rho)| \leqslant \left\| H(\Lambda_{n+1}) - \sum_{j=1}^{2^d} H(\Lambda_{n,j})\right\|.$$

On dividing this inequality by $|\Lambda_{n+1}|$, and using (C.1), we see from this inequality that

$$\left|\hat{E}_{\Lambda_{n+1}}(\rho)/|\Lambda_{n+1}| - \hat{E}_{\Lambda_n}(\rho)/|\Lambda_n|\right| < c/2^{nd\varepsilon}.$$

Hence, as the sum of the geometric progression $\{2^{-nd\varepsilon}\}$ converges, so too does $\hat{E}_{\Lambda_n}(\rho)/|\Lambda_n|$, uniformly in ρ. Thus, denoting the limit of this energy density by $\hat{e}(\rho)$,

$$\hat{E}_{\Lambda_n}(\rho)/|\Lambda_n| \to \hat{e}(\rho), \quad \text{uniformly w.r.t. } \rho, \text{ as } n \to \infty. \tag{C.3}$$

Convergence of $\hat{S}_\Lambda(\rho)/|\Lambda|$. From the subadditivity of entropy it follows that

$$\hat{S}_{\Lambda_{n+1}}(\rho) \leqslant \sum_{j=1}^{2^d} \hat{S}_{\Lambda_{n,j}}(\rho),$$

and therefore, for translationally invariant states ρ,

$$\hat{S}_{\Lambda_{n+1}}(\rho) \leqslant 2^d \hat{S}_{\Lambda_n}(\rho).$$

On dividing this inequality by $|\Lambda_{n+1}|$, we see that

$$\hat{S}_{\Lambda_{n+1}}(\rho)/|\Lambda_{n+1}| \leqslant \hat{S}_{\Lambda_n}(\rho)/|\Lambda_n|,$$

i.e. that the entropy density $\hat{S}_{\Lambda_n}(\rho)/|\Lambda_n|$ decreases monotonically with n. Therefore since, by (25), this quantity is non-negative and, consequently, lower-bounded, it converges to its infinium, $\hat{s}(\rho)$, as $n \to \infty$, i.e.

$$\hat{s}(\rho) = \lim_{n\to\infty} \hat{S}_{\Lambda_n}(\rho)/|\Lambda_n| = \inf_n \hat{S}_{\Lambda_n}(\rho)/|\Lambda_n|. \tag{C.4}$$

Affine property of $\hat{e}$ *and* $\hat{s}$. It follows immediately from (24) and (29) (or (C.3)) that $\hat{E}_\Lambda(\rho)$ and thus $\hat{e}(\rho)$ is affine.

The key to the proof that $\hat{s}$ is also affine is that, on the one hand, $x \ln x$ is a convex function of the positive real variable x, i.e.

$$(\lambda x_1 + (1-\lambda)x_2)\ln(\lambda x_1 + (1-\lambda)x_2) \leqslant \lambda x_1 \ln x_1 + (1-\lambda)x_2 \ln x_2$$

for $0 < \lambda < 1$; while, on the other hand, this convexity is limited to the extent that

$$\begin{aligned}&(\lambda x_1 + (1-\lambda)x_2)\ln(\lambda x_1 + (1-\lambda)x_2)\\ &\qquad \geqslant \lambda x_1 \ln x_1 + (1-\lambda)x_2 \ln x_2 + x_1\lambda \ln \lambda + x_2(1-\lambda)\ln(1-\lambda).\end{aligned}$$

These two inequalities have been proved by Lanford and Robinson (1968*a*) to have the following operator-theoretic generalizations for the local entropy functional $\hat{S}_\Lambda(\rho)(\equiv -k\,\mathrm{Tr}_\Lambda(\rho_\Lambda \ln \rho_\Lambda))$:

$$\hat{S}_\Lambda(\lambda\rho_1 + (1-\lambda)\rho_2) \geqslant \lambda\hat{S}_\Lambda(\rho_1) + (1-\lambda)\hat{S}_\Lambda(\rho_2),$$

and

$$\hat{S}_\Lambda(\lambda\rho_1 + (1-\lambda)\rho_2) \leqslant \lambda\hat{S}_\Lambda(\rho_1) + (1-\lambda)\hat{S}_\Lambda(\rho_2) - k(\lambda \ln \lambda + (1-\lambda)\ln(1-\lambda)).$$

On dividing these inequalities by $|\Lambda|$ and passing to the infinite volume limit, we see that

$$\hat{s}(\lambda\rho_1 + (1-\lambda)\rho_2) \geqslant \text{ and } \leqslant \lambda\hat{s}(\rho_1) + (1-\lambda)\hat{s}(\rho_2),$$

which means that $\hat{s}$ is affine.

Continuity of $\hat{e}(\rho)$. This follows immediately from the continuity of $\hat{E}_\Lambda(\rho)$ and the fact that, by (C.3), $\hat{e}(\rho)$ is the uniform limit of $\hat{E}_{\Lambda_n}(\rho)/|\Lambda|$.

Upper semicontinuity of $\hat{s}(\rho)$. Suppose that $\{\rho_l\}$ is a sequence of translationally invariant states such that $\rho_l \to \bar{\rho}$ and $\hat{s}(\rho_l) \to \bar{s}$ as $l \to \infty$. Then, as $\hat{s}(\rho)$ is a lower bound of $\hat{S}_{\Lambda_n}(\rho)/|\Lambda_n|$, by (C.4),

$$\hat{s}(\rho_l) \leqslant \hat{S}_{\Lambda_n}(\rho_l)/|\Lambda_n|.$$

Therefore, in view of the continuity of $\hat{S}_{\Lambda_n}$, the limiting form of this inequality, as $l \to \infty$, is

$$\bar{s} \leqslant \hat{S}_{\Lambda_n}(\bar{\rho})/|\Lambda_n|.$$

On passing to the limit $n \to \infty$, this becomes

$$\bar{s} \leqslant \hat{s}(\bar{\rho}),$$

which means that $\hat{s}$ is upper semicontinuous.

Appendix D
Proof that translationally invariant LTS states are GTS in the case of finite-range interactions

Let us assume the interactions to be of finite range. Then, if Λ and Λ' are bounded regions, with Λ' containing Λ, the interaction energy $W(\Lambda, \Lambda' \backslash \Lambda)$ between the particles in Λ and those in $\Lambda' \backslash \Lambda$ becomes independent of Λ whenever the Euclidean distance between the boundaries of Λ and Λ' exceeds a certain value, R. Thus, denoting the value of this Λ'-independent energy by $W(\Lambda)$, it follows from (18) that the local Hamiltonians for the regions Λ, Λ', and $\Lambda'\backslash\Lambda$ are related by the formula

$$H(\Lambda') = H(\Lambda) + H(\Lambda'\backslash\Lambda) + W(\Lambda), \tag{D.1}$$

for Λ' sufficiently large. Furthermore, it follows rather easily† from the finite range of the interactions that $W(\Lambda)$ behaves as a surface effect, in the sense that

$$\lim_{\Lambda\uparrow} \|W(\Lambda)\|/|\Lambda| = 0. \tag{D.2}$$

Suppose now that ρ is a translationally invariant LTS state. Then, if ρ' is any other state of Σ which coincides with ρ' outside some bounded region Λ, it follows from the local stability of ρ that

$$\Delta\hat{F}(\rho \mid \rho') \geqslant 0.$$

In view of eqns (40)–(42) and (D.1), this inequality is equivalent to

$$\hat{E}_\Lambda(\rho') + \rho'(W(\Lambda)) \geqslant \hat{E}_\Lambda(\rho) + \rho(W(\Lambda)) + T\Delta\hat{S}(\rho \mid \rho'), \tag{D.3}$$

where

$$\Delta\hat{S}(\rho \mid \rho') = \lim_{\Lambda'\uparrow} (S_{\Lambda'}(\rho') - S_{\Lambda'}(\rho)). \tag{D.4}$$

We choose ρ' so that it reduces to the canonical state $\bar{\rho}_\Lambda$, corresponding to the density matrix $e^{-\beta H(\Lambda)}/\mathrm{Tr(idem)}$, inside Λ, carries no correlations between the observables inside and outside Λ, and, of course, coincides with ρ outside Λ. Thus, for $\Lambda' \supset \Lambda$,

$$\rho'_{\Lambda'} = \bar{\rho}_\Lambda \otimes \rho_{\Lambda'\backslash\Lambda}. \tag{D.5}$$

From this product form of $\rho'_{\Lambda'}$, it follows that

$$\hat{S}_{\Lambda'}(\rho') = \hat{S}_\Lambda(\rho') + S_{\Lambda\backslash\Lambda'}(\rho),$$

and consequently, by (D.4), that

$$\Delta\hat{S}(\rho \mid \rho') = \hat{S}_\Lambda(\rho') - \lim_{\Lambda'\uparrow} (\hat{S}_{\Lambda'}(\rho) - \hat{S}_{\Lambda'\backslash\Lambda}(\rho)).$$

By the subadditivity of entropy, this implies that

$$\Delta\hat{S}(\rho \mid \rho') \geqslant \hat{S}_\Lambda(\rho') - S_\Lambda(\rho),$$

and from this inequality and (D.3) it follows that

$$\hat{E}_\Lambda(\rho') - T\hat{S}_\Lambda(\rho') + \rho'(W(\Lambda)) \geqslant \hat{E}_\Lambda(\rho) - T\hat{S}_\Lambda(\rho) + \rho(W(\Lambda)),$$

i.e.

$$\hat{F}_\Lambda(\rho') \geqslant \hat{F}_\Lambda(\rho) + \rho(W(\Lambda)) - \rho'(W(\Lambda)). \tag{D.6}$$

† For, if $\partial_1\Lambda$ and $\partial_2\Lambda$ are the 'shells' consisting of the sites inside and outside Λ, respectively, which lie within R of the surface of that region, then $W(\Lambda) = W(\Lambda, \partial_2\Lambda) = W(\partial_2\Lambda, \partial_1\Lambda)$; and consequently by (18) and (20), $\|W(\Lambda)\| < C(|\partial_1\Lambda| + |\partial_2\Lambda|)$. Since both $|\partial_1\Lambda|$ and $|\partial_2\Lambda|$ are of the order of $R \times$ the surface area of Λ, it follows that (D.2) is valid.

Furthermore, since ρ' reduces to the canonical state $\bar{\rho}_{\Lambda}$ in Λ, by (D.5), $\hat{F}_{\Lambda}(\rho')$ is simply the canonical free energy $\Phi_{\Lambda}(T)$. Therefore, as the moduli of the last two terms in (D.6) are both less than, or equal to, $\|W(\Lambda)\|$, it follows from that inequality that

$$\Phi_{\Lambda}(T) \geqslant \hat{F}_{\Lambda}(\rho) - 2\,\|W(\Lambda)\|.$$

On dividing this formula by $|\Lambda|$ and passing to the infinite volume limit, we see that, in view of the fact that $W(\Lambda)$ behaves as a surface effect, by (D.2),

$$\phi(T) \geqslant \hat{f}(\rho).$$

Since $\phi(T)$ is the lower bound of the functional $\hat{f}$, the equality sign is operative here, and so $\phi(T) = \hat{f}(\rho)$, which means that ρ is GTS, as required.

4

Thermodynamics and phase structure

4.1. Introduction

The statistical mechanical description of equilibrium states in the previous chapter was expressed in terms of the quantum mechanical structure of infinite systems. Classical thermodynamics, on the other hand, is generally expressed in terms of macroscopic variables, given by *extensive conserved quantities*,† such as energy, momentum and magnetic moment. Our object now is to pass from the quantum statistical to the classical thermodynamical description of equilibrium states and thereby to construct a framework for the theory of phases.

We start by recalling‡ that the equilibrium states of an infinite system are those that maximize the entropy density, given the value of the energy density. In general, there may be several such states corresponding to an energy density e or, equivalently, to a temperature T, and the question arises as to which variables may be used to distinguish between them. In simple concrete cases, it is clear what these variables are. For example, in the case of an Ising ferromagnet, the different equilibrium states at the same energy density, or temperature, are completely specified by the values of their polarizations.§ More generally, both classical¶ and statistical‖ thermodynamics are dependent on the assumption that, for any macroscopic system, there is some set $(Q_1, \ldots, Q_n)$ of extensive conserved quantities, whose densities serve to specify the equilibrium states for given energy density. We shall carry this assumption over to the theory of infinite systems. Specifically, we shall assume that, for an infinite system Σ, there is a set of extensive conserved quantities comprising the energy, E, and a finite number, n, of others $(Q_1, \ldots, Q_n) \equiv Q$, which form a complete macroscopic description of the equilibrium states in the following sense: *given the values of the global densities* (e, q) *of* (E, Q), *there is precisely one equilibrium state*, ρ, *i.e. one state that maximizes the entropy density of the system*. This implies that, if there exist any other extensive conserved quantities, they are redundant in the sense that their equilibrium densities are functions of (e, q).

† We shall give a precise definition of the term 'extensive conserved quantity' in §4.2.
‡ See Note at the end of Chapter 3.
§ See Messager and Miracle-Sole (1975).
¶ See Callen (1960).
‖ See Landau and Lifshitz (1959), §4.

In view of our above assumption, we define the *macroscopic states* of Σ by the values of the global intensive variables (e, q). The equilibrium entropy density is then the function, $s(e, q)$, of these variables given by the maximum value of the functional $\hat{s}(\rho)$, subject to the constraint that the expectation values of the densities of (E, Q) are (e, q). The other thermodynamic functions may be defined in terms of the entropy function, $s(e, q)$, by standard procedures. For example, the Gibbs potential is the function $\phi(y, T)$ of the intensive variables $y \equiv (y_1, \ldots, y_n)$, conjugate to Q, and the temperature T, given by the formula

$$\phi(y, T) = \min_{e,q} (e - Ts(e, q) - y \cdot q).$$

We remark here that the variables y correspond to external fields, or chemical potentials, and thus that (y, T) may be regarded as 'control variables'. It is evident from the above definition that the Gibbs potential $\phi(y, T)$ reduces to the canonical free energy density, $\phi(T)$, of the previous chapter, when $y = 0$.

The phase structure of Σ is determined by the dependence of both its thermodynamic potentials and its equilibrium states on the control variables y and T. The picture of this structure given by Classical Thermodynamics is that the (y, T)-space is divided into regions I, II, ... in which the Gibbs potential $\phi(y, T)$ is analytic, and that these regions are separated from one another by surfaces on which this function is singular (cf. Fig. 5). In this picture, the domains I, II, ... correspond to the single phase regions, and the surfaces separating them to two- or multi-phase regions.

On the other hand, in the statistical mechanical picture of the previous

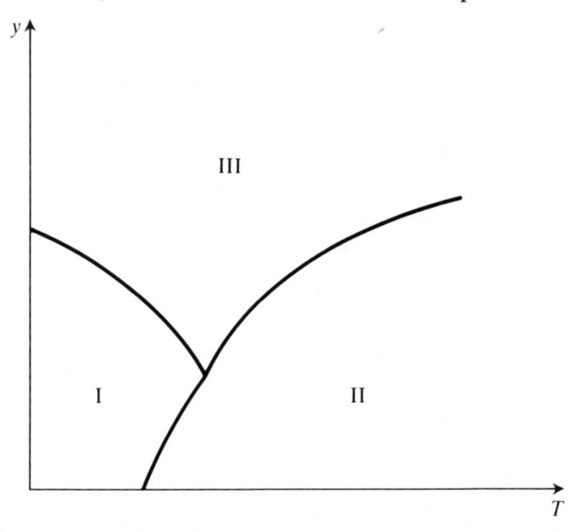

Fig. 5. Example of a phase diagram. I, II, and III are the single-phase regions, and the boundaries separating them are the two- or multi-phase regions.

chapter, the single-phase region consists of the points (y, T) for which there is just one equilibrium state, and the multi-phase region consists of the points (y, T) for which there is macroscopic degeneracy. The question now arises as to whether this picture of phase structure is consistent with the classical thermodynamical one. In fact, we shall prove (§4.5) that it is, and specifically that *macroscopic degeneracy occurs at just those points* (y, T) *where the Gibbs potential has a discontinuous derivative*. As a consequence of this result, we are able to construct a framework for the theory of phases that is consistent with both the classical thermodynamical and quantum statistical pictures.

We shall present our formulation of the thermodynamics and phase structure of Σ as follows. In §4.2, we shall specify precisely what is meant by the above assumption that (E, Q) form a complete set of extensive conserved quantities. In §4.3, we shall give the statistical mechanical definitions of the thermodynamic potentials, and shall obtain some of their key mathematical properties, especially convexity (or concavity), continuity, and analyticity. In §4.4, we shall formulate the equilibrium states corresponding to arbitrary (y, T) in macroscopic and microscopic terms. This will lead to a characterization of *pure phases*, due originally to Ruelle (1967), as equilibrium states that cannot be resolved into mixtures of other equilibrium states. Moreover, we shall show that this is equivalent to the classical thermodynamical characterization of a pure phase by the condition that it carries no dispersion in the densities of the extensive conserved quantities. In §4.5, we shall establish the connection between thermodynamical discontinuity and macroscopic degeneracy specified in the previous paragraph. In §4.6, we shall discuss the consequences of this result for the theory of the phase structure of matter, with particular reference to the classifications of phase transitions due to Ehrenfest and Landau. In §4.7, we shall derive the classical thermodynamical relationship,

$$T\,\mathrm{d}s = \mathrm{d}e + y\,.\,\mathrm{d}q,$$

for a pure phase of an *infinite* system, and also the Clausius–Clapeyron equation, from the convexity and continuity properties of the thermodynamic potentials. In §4.8, we shall summarize the principal conclusions of this chapter. There will be two Appendices, devoted to proofs of some mathematical points made in the chapter.

For simplicity, we shall confine our explicit constructions here to the lattice model of the previous chapter: the generalization to the continuous systems, for which the global stability conditions are well-defined,† presents no essential difficulties.

† See §3.8 for a brief discussion of those conditions for continuous systems.

4.2. Extensive conserved quantities

For a *finite* system, S, an extensive conserved quantity is an observable, $Q(S)$, which commutes with the Hamiltonian and which possesses the property that, if S is the composite of spatially disjoint subsystems, S_1 and S_2, then $Q(S)=Q(S_1)+Q(S_2)$, with perhaps some correction due to 'surface effects'.

Correspondingly, we define an extensive conserved quantity for the *infinite* lattice system, Σ, to be a family $\{Q(\Lambda)\}$ of observables for the respective bounded regions $\{\Lambda\}$, such that

(Q1) $Q(\Lambda)$ commutes with $H(\Lambda)$;

(Q2) $Q(\Lambda)$ transforms covariantly with respect to space translations, i.e.

$$\gamma(x)Q(\Lambda)=Q(\Lambda+x);$$

(Q3) $Q(\Lambda)$ is additive in Λ, up to a surface effect, in the sense that

$$\|Q(\Lambda_1\cup\Lambda_2)-Q(\Lambda_1)-Q(\Lambda_2)\|<\eta(\Lambda_1)+\eta(\Lambda_2)$$

for disjoint Λ_1, Λ_2, where $\lim_{\Lambda\uparrow}\eta(\Lambda)/|\Lambda|^{1-\alpha}=0$ for some α between 0 and 1; and

(Q4) the density of Q is uniformly bounded, in the sense that

$$\|Q(\Lambda)\|<c_Q\,|\Lambda|$$

for all bounded regions Λ, where c_Q is some finite constant.

One sees immediately from our specifications of §3.3 that the energy of the system meets our conditions for an extensive conserved quantity, since the local Hamiltonians $\{H(\Lambda)\}$ satisfy (Q.1–4). One also sees that the covariance and extensivity properties (Q2) and (Q3) enable us to define the global intensive variable, corresponding to the density of Q, analogously with our definition of energy density in §3.4. Thus, we define the density of Q to be the functional, $\hat{q}$, of the translationally invariant states by the formula

$$\hat{q}(\rho)=\lim_{\Lambda\uparrow}\rho(Q(\Lambda))/|\Lambda|. \tag{1}$$

Furthermore, this functional, like the energy density $\hat{e}$, is affine and continuous.†

We now introduce the following fundamental assumption, whose basis was discussed in §4.1.

(A)‡ *There is a finite set* $(E, Q_1, \ldots, Q_n)$ *of extensive conserved quantities*,

† The methods employed in Appendix C of Chapter 3 to establish these properties for $\hat{e}$ may be carried over to prove the same for $\hat{q}$.

‡ The maximization of $\hat{s}$ by at least one state, subject to the given constraints, is guaranteed by the upper semicontinuity of $\hat{s}$, which was established in Chapter 3.

with E the energy, such that, if the densities $(\hat{e}, \hat{q}_1, \ldots, \hat{q}_n)$ *of these quantities are constrained to take values* $(e, q_1, \ldots, q_n)$, *say, then the entropy functional* $\hat{s}(\rho)$ *is maximized by precisely one translationally invariant state.*

Under this assumption, we term $(E, Q_1, \ldots, Q_n)$ a *complete* set of extensive conserved quantities. We shall always take it that this set is minimal, in the sense that it contains no subsets that are complete.

Denoting the sets $(Q_1, \ldots, Q_n)$, $(\hat{q}_1, \ldots, \hat{q}_n)$ and $(q_1, \ldots, q_n)$ by $Q, \hat{q}$, and q, respectively, we define $\bar{\rho}(e, q)$ to be the (unique) state that maximizes the entropy functional $\hat{s}(\rho)$, subject to the condition that the densities of (E, Q) takes the values (e, q); and we define $s(e, q)$ to be the entropy density of this state, i.e.

$$s(e, q) = \hat{s}(\bar{\rho}(e, q)) = \max\{\hat{s}(\rho) \mid \hat{e}(\rho) = e; \hat{q}(\rho) = q\}. \tag{2}$$

Evidently, the domain of definition, D, of the function $s(e, q)$, and of $\bar{\rho}(e, q)$, is restricted to the points (e, q) lying within the range of the functionals $(\hat{e}(\rho), \hat{q}(\rho))$ as ρ runs through the translationally invariant states. As we shall show in Appendix A, it follows from the properties established for the functionals $\hat{e}$, $\hat{q}$, and $\hat{s}$ that D is a *closed convex* region in $(n+1)$-dimensional Euclidean space, and that the function $s(e, q)$ is *concave and upper semicontinuous*.

4.3. The Helmholtz and Gibbs potentials

We now define the Helmholtz and Gibbs potentials in terms of the entropy function $s(e, q)$ by standard thermodynamical formulae.†
Thus, we define the *Helmholtz potential* (or free energy density) to be the function of the temperature T and the global intensive variable q, given by the formula‡

$$\psi(q, T) = \min_e (e - Ts(e, q)); \tag{3}$$

and we define the *Gibbs potential* to be the function of the intensive variables y, conjugate to Q, and the temperature T, given by

$$\phi(y, T) = \min_{e,q} (e - Ts(e, q) - y \cdot q). \tag{4}$$

Equivalently, by (2), we may express these formulae for the Helmholtz and Gibbs potentials in the following quantum statistical forms

$$\psi(q, T) = \min\{\hat{f}(\rho) \equiv \hat{e}(\rho) - T\hat{s}(\rho) \mid \hat{q}(\rho) = q\}, \tag{5}$$

† See Callen (1960).

‡ The minimizability of $(e - Ts(e, q))$ and of $(e - Ts(e, q) - y \cdot q)$ in eqns (3) and (4), respectively, is guaranteed by the facts that $s(e, q)$ is upper semicontinuous and that its domain of definition is closed (cf. argument in §3.2.3).

and

$$\phi(y, T) = \min_{\rho} (\hat{f}(\rho) - y \,.\, \hat{q}(\rho)) \equiv \min_{\rho} (\hat{e}(\rho) - T\hat{s}(\rho) - y \,.\, \hat{q}(\rho)). \quad (6)$$

On comparing this last equation with the formula (31) of Chapter 3, for the canonical free energy density, $\phi(T)$, we see that $\phi(y, T)$ is given by replacing $H(\Lambda)$ by $H(\Lambda) - y \,.\, Q(\Lambda)$ in the eqns (32) and (33) there, i.e.

$$\phi(y, T) = \lim_{\Lambda\uparrow} |\Lambda|^{-1}\Phi_{\Lambda}(y, T), \quad (7)$$

where

$$\Phi_{\Lambda}(y, T) = -kT \ln \mathrm{Tr}_{\Lambda} \exp - \beta(H(\Lambda) - y \,.\, Q(\Lambda)), \quad (8)$$

which is the Gibbs free energy of a finite system of the given species occupying the region Λ.

Let us now note some key mathematical properties of the thermodynamic potentials.

Analyticity, singularities. It follows from the formula (8) that, in view of the analyticity of the exponential function and the finite dimensionality† of $\mathcal{H}(\Lambda)$, the potential $\Phi_{\Lambda}(y, T)$ must be analytic in all its arguments for real $y_1, \ldots, y_n$ and positive T. This signifies that the Gibbs potential of a *finite* system cannot have any singularities.

On the other hand, since the limit of a sequence of analytic functions is not necessarily analytic, it follows from eqn (7) that the Gibbs potential $\phi(y, T)$ for the infinite system Σ *might* have singularities. In fact, it has been proved that the Gibbs potentials of various tractable models‡ do indeed have them. Thus we see that the passage to the thermodynamical limit opens up the way for a characterization of *phase transitions* by singularities in the Gibbs potential.

Convexity and concavity properties. As we shall show in Appendix A, it follows from the definitions of the Helmholtz and Gibbs potentials, together with the concavity of $s(e, q)$, noted at the end of §4.2, that

(a) $\psi(q, T)$ *is convex in* q *and concave in* T, i.e.

$$\psi(\lambda q + (1 - \lambda)q', T) \leq \lambda\psi(q, T) + (1 - \lambda)\psi(q', T) \quad (9)$$

and

$$\psi(q, \lambda T + (1 - \lambda)T') \geq \lambda\psi(q, T) + (1 - \lambda)\psi(q, T'), \quad (10)$$

† In fact, the analyticity of Φ_{Λ} may also be established in cases where $\mathcal{H}(\Lambda)$ is infinite-dimensional, as in finite continuous systems. For, granted the finiteness of $\Phi_{\Lambda}(y, T)$, and hence the absolute convergence of $\mathrm{Tr}_{\Lambda}\, e^{-\beta(H(\Lambda) - y \,.\, Q(\Lambda))}$ for *all* positive T and a range of real values of y, it is a straightforward matter to deduce that Φ_{Λ} is analytic there.

‡ Good examples are provided by Onsager's (1944) solution of the Ising model; and by the treatments by Ginibre (1969), Robinson (1969), and Dyson, Lieb, and Simon (1978) of various ferromagnetic models.

for $0<\lambda<1$; and

(b) $\phi(y, T)$ *is concave*, i.e.

$$\phi(\lambda y+(1-\lambda)y', \lambda T+(1-\lambda)T') \geqslant \lambda\phi(y, T)+(1-\lambda)\phi(y', T'), \quad (11)$$

for $0<\lambda<1$.

These properties impose severe restrictions on the behaviour of the thermodynamic potentials. For, as noted in the discussion of §3.2.1, a convex function cannot have a local, non-absolute, minimum, while neither a convex nor a concave function can support a discontinuity in the interior of the range of its argument, though it might have a discontinuous derivative there. Hence we may draw the following conclusions from the above properties (a), (b) of the Helmholtz and Gibbs potentials.

(1) For any fixed T, the function $\psi(q, T)$ cannot have a local, non-absolute minimum. Thus, *one cannot hope to characterize metastable states by local, non-absolute minima of the Helmholtz potential*.

(2) The Gibbs potential $\phi(y, T)$ is continuous for all real $y_1, \ldots, y_n$ and positive T, though it might have discontinuous derivatives there. Hence, *singularities in the Gibbs potential may be manifested by discontinuities in derivatives of ϕ, but not in ϕ itself*. This has important consequences for phase transitions.

4.4. Equilibrium states, pure phases,† and mixtures

The free energy density functional for Σ, corresponding to temperature T and external fields (or chemical potentials) y, is $\hat{f}(\rho)-y\,.\,\hat{q}(\rho)$. Thus, at the *quantum mechanical* level, the equilibrium states corresponding to the thermodynamical conditions given by (y, T) are characterized by the global stability condition that they minimize this functional. They therefore comprise the set, μ_{yT}, of states ρ for which

$$\phi(y, T)=\hat{f}(\rho)-y\,.\,\hat{q}(\rho) \equiv \hat{e}(\rho)-T\hat{s}(\rho)-y\,.\,\hat{q}(\rho). \quad (12)$$

At the *macroscopic* level, on the other hand, the equilibrium states (e, q) are those that minimize the free energy density $e-Ts(e, q)-y\,.\,q$. Thus, by (4), the equilibrium macrostates (e, q) corresponding to (y, T) are characterized by the property that

$$\phi(y, T)=e-Ts(e, q)-y\,.\,q. \quad (13)$$

We denote this set of equilibrium macrostates by M_{yT}. We now observe that, since by our basic assumption (A), there is precisely one (micro-) state $\bar{\rho}(e, q)$ of Σ that maximizes the entropy functional $\hat{s}$, subject to the

† The characterization of pure phases suggested in §2.4.4 reduces to that presented here for the case of translationally invariant states with short-range correlations.

constraint that the macrostate is (e, q), it follows from the definitions (12) and (13) that *the sets, μ_{yT} and M_{yT}, of equilibrium microstates and macrostates are in one-to-one correspondence*, i.e.

(B) *μ_{yT} is the set of states $\bar{\rho}(e, q)$ for which (e, q) belongs to M_{yT}.*

In order to relate equilibrium states to thermodynamic phases, we first note that the sets μ_{yT} and M_{yT} are both *convex*, for the following reasons. Since the functionals $\hat{f}$ and $\hat{q}$ are affine, it follows that if ρ_1 and ρ_2 satisfy the microscopic equilibrium condition (13), then so too does $\lambda\rho_1 + (1 - \lambda)\rho_2$, for $0 < \lambda < 1$, which means that μ_{yT} is convex. Further, if the global densities of (E, Q) take the values (e_1, q_1) and (e_2, q_2) respectively, for these equilibrium states ρ_1 and ρ_2, then their values for the state $\lambda\rho_1 + (1 - \lambda)\rho_2$ are $(\lambda e_1 + (1 - \lambda)e_2, \lambda q_1 + (1 - \lambda)q_2)$. In view of the above correspondence (B) between the microscopic and macroscopic descriptions of the equilibrium states, this signifies that, if (e_1, q_1) and (e_2, q_2) belong to M_{yT}, then so too does $(\lambda e_1 + (1 - \lambda)e_2, \lambda q_1 + (1 - \lambda)q_2)$, which means that M_{yT} is convex. It also implies that the correspondence between μ_{yT} and M_{yT} is an *isomorphism,* i.e. that $\bar{\rho}(\lambda e_1 + (1 - \lambda)e_2, \lambda q_1 + (1 - \lambda)q_2) = \lambda\bar{\rho}(e_1, q_1) + (1 - \lambda)\bar{\rho}(e_2, q_2)$ when $(e_1 . q_1)$ and (e_2, q_2) belong to M_{yT}.

The relevance of the convexity of μ_{yT}, and m_{yT}, to the phase structure of Σ emerges when one takes account of the classic Krein–Millman theorem† on convex sets, which tells us that

(a) μ_{yT} contains so-called *extremal elements*, characterized by the property that they cannot be expressed as mixtures, $\lambda\rho_1 + (1 - \lambda)\rho_2$, of different elements of this set; and
(b) any state in μ_{yT} is a weighted sum, or integral, of the extremals.

The extremals are therefore the 'building blocks' of μ_{yT}. This suggests that, as first proposed by Ruelle (1965), the extremals should be considered as *pure thermodynamical phases*, and the other equilibrium states as mixtures. The question naturally arises as to whether this accords with the *classical thermodynamical* characterization of pure phases by the property that they carry no dispersions in the global densities of the extensive conserved quantities (E, Q). In fact, it does so, for the following reasons.

As proved‡ by Ruelle (1967, 1969a), in support of his idea that the

† See Yosida (1965), p. 362. A simple version of this theorem is that, if Δ is a closed bounded convex region of a Euclidean space, say, then any point in Δ is the centroid of some distribution of points on its boundary. More generally the theorem states that any point in a convex compact region of a vector space is a weighted sum of its extremal points. The applicability of the theorem to the equilibrium states of Σ thus depends on the compactness property of the states, established in Chapter 3.

‡ The proof is based on a version of ergodic theory pertaining to *spatial*, rather than temporal, translations. Here, the indecomposability of the extremals into mixtures of other equilibrium states corresponds to the metrical indecomposability of energy surfaces in classical ergodic theory.

extremals correspond to pure phases, these are the only equilibrium states which carry no dispersions in the global space averages of local observables. In other words, if ρ is an equilibrium state and $\tilde{A}_\Lambda$ is the space average of a local observable A over a bounded region Λ, i.e.

$$\tilde{A}_\Lambda = |\Lambda|^{-1} \sum_{x \in \Lambda} A(x), \quad \text{where } A(x) \equiv \gamma(x)A, \tag{14}$$

then

$$\lim_{\Lambda\uparrow}[\rho(\tilde{A}_\Lambda^2) - \rho(\tilde{A}_\Lambda)^2] = 0, \tag{15}$$

for all local observables A, if and only if ρ is extremal. An extension of this result, which we shall prove in Appendix B, is that the global densities of the extensive conserved quantities are also dispersion-free in the extremal equilibrium states, i.e. that (15) remains valid for those states if $\tilde{A}_\Lambda$ is replaced by $H(\Lambda)/|\Lambda|$ or $Q(\Lambda)/|\Lambda|$. Hence, in view of the isomorphism between the equilibrium microstates, μ_{yT}, and macrostates, M_{yT}, we infer that

(1) if $\bar{\rho}(e, q)$ is an extremal equilibrium state, i.e. if (e, q) is an extremal element of M_{yT}, then the global densities of (E, Q) take the precise values (e, q) for that state; and
(2) if ρ is a non-extremal equilibrium state, and is thus a statistical mixture of extremals $\bar{\rho}(e, q)$, then the global densities of (E, Q) are given by the same mixture of the macrostates (e, q), and thus carry dispersions.

In other words, *the extremals are the equilibrium states that satisfy the classical thermodynamical condition for pure phases*, namely that they carry no dispersions in the global densities of the extensive conserved variables (E, Q). We take the view that this result substantiates the quantum statistical argument for regarding the extremals as pure phases, as they cannot be resolved into mixtures of other equilibrium states. Thus, we shall henceforth assume that *the extremal elements of μ_{yT}, or equivalently of M_{yT}, represent the pure thermodynamical phases corresponding to (y, T)*.

4.5. Macroscopic degeneracy and thermodynamic discontinuity

Since it is known empirically that the coexistence of different phases of a system is often accompanied by a singularity in a thermodynamical potential, it is natural to enquire whether there is any fundamental connection between macroscopic degeneracies and thermodynamic singularities. In fact, we shall now show that there is, in that *the Gibbs potential $\phi(y, T)$ has a discontinuous first derivative at (y, T) if and only if there is a macroscopic degeneracy corresponding to these values of the thermodynamic parameters*. This signifies that, as is generally assumed in Classical Thermodynamics, the region of (y, T)-space where the Gibbs

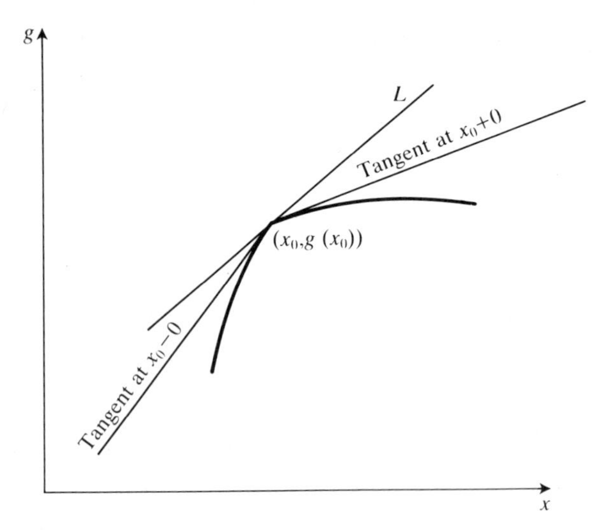

Fig. 6. Graph of concave function with possibly discontinuous derivative at x_0.

potential is differentiable with respect to all its arguments is indeed the one-phase region.

A mathematical key to this result is that the graph of a convex, or concave, function has one or more tangents (or tangent planes) at a point according to whether or not the function is differentiable there (cf. Fig. 6). In fact, we shall prove the connection between macroscopic degeneracy and thermodynamical discontinuity by obtaining a correspondence† between the equilibrium macrostates at (y, T) and the tangent planes to the graph of the Gibbs potential there. Specifically, we shall show that the system has one or more equilibrium macrostates at (y, T) according to whether or not there is only tangent plane to the graph of $\phi(y, T)$ at that point. In view of the isomorphism between the equilibrium macrostates, M_{yT}, and microstates, μ_{yT}, this means that the differentiability of the Gibbs potential at (y, T) constitutes the necessary and sufficient condition for the uniqueness of the equilibrium state there; i.e. it establishes the connection, stated above, between macroscopic degeneracy and thermodynamic discontinuity.

As a preliminary to our proof of this result we note the following standard properties of an arbitrary concave function‡ $g(x)$ of one variable, as illustrated by Fig. 6.

(C.1) At any value, x_0, of x, g has right and left derivatives $\dfrac{d^{(\pm)}}{dx} g(x_0)$,

† This correspondence is analogous to that obtained by Lanford and Robinson (1968*b*) between the equilibrium microstates and the tangent planes to the graph of a thermodynamic potential, expressed as a function of an infinite set of micro-variables.

‡ See Roberts and Varberg (1973), p. 7.

defined by the formula

$$\frac{\mathrm{d}^{(\pm)}}{\mathrm{d}x} g(x_0) = \lim_{h \to \pm 0} \frac{g(x_0 + h) - g(x)}{h}.$$

Furthermore,

$$\frac{\mathrm{d}^{(+)}}{\mathrm{d}x} g(x_0) \leqslant \frac{\mathrm{d}^{(-)}}{\mathrm{d}x} g(x_0),$$

equality occurring if and only if g is differentiable at x_0.

(C.2) If a straight line L, of slope k, through the point $(x_0, g(x_0))$, lies above the graph of g, i.e. if

$$g(x_0 + h) - g(x_0) \leqslant hk \quad \text{for all real } h,$$

then

$$\frac{\mathrm{d}^{(+)}}{\mathrm{d}x} g(x_0) \leqslant k \leqslant \frac{\mathrm{d}^{(-)}}{\mathrm{d}x} g(x_0).$$

Thus, if $g(x)$ is differentiable at x_0, then $k = \dfrac{\mathrm{d}g}{\mathrm{d}x}(x_0)$.

(C.3) There exist sequences $\{x_n^{(+)}\}$ and $\{x_n^{(-)}\}$, converging to x_0 from above and below, respectively, such that $g(x)$ is differentiable at the points $\{x_n^{(\pm)}\}$ and

$$\frac{\mathrm{d}^{(\pm)}}{\mathrm{d}x} g(x_0) = \lim_{n \to \infty} \frac{\mathrm{d}g}{\mathrm{d}x}(x_n^{(\pm)}).$$

We shall now utilize these properties of concave functions in order to relate the equilibrium states of Σ to the derivatives of $\phi(y, T)$ and to establish the connection between macroscopic degeneracy and discontinuity in these derivatives. We start by recalling that the equilibrium macrostates (e, q) corresponding to the values (y, T) of the control variables are characterized by the condition (13), i.e.

$$\phi(y, T) = e - Ts(e, q) - y \,.\, q. \tag{16}$$

On the other hand, as the Gibbs potential $\phi(y', T')$, for arbitrary (y', T') is the minimum value of $e' - T's(e', q') - y' \,.\, q'$, as (e', q') runs through the macrostates, it follows that

$$\phi(y', T') \leqslant e - T's(e, q) - y', q;$$

and therefore, in view of the previous equation

$$\phi(y', T') - \phi(y, T) \leqslant -(T' - T)s(e, q) - (y' - y) \,.\, q.$$

Hence, by (C.2), as $\phi(y, T)$ is concave in each of its arguments,

$$\frac{\partial^{(+)}}{\partial T}\phi(y, T) \leqslant -s(e, q) \leqslant -\frac{\partial^{(-)}}{\partial T}\phi(y, T), \tag{17}$$

and

$$\frac{\partial^{(+)}}{\partial y_j}\phi(y, T) \leqslant -q_j \leqslant \frac{\partial^{(-)}}{\partial y_j}\phi(y, T), \text{ for } j = 1, \ldots, n. \tag{18}$$

In the case where ϕ is differentiable with respect to all its arguments at (y, T), these inequalities imply that

$$s(e, q) = -\frac{\partial \phi}{\partial T}(y, T) \quad \text{and} \quad q = -\frac{\partial \phi}{\partial y}(y, T), \tag{19}$$

and consequently, by (13), that e and q are *uniquely determined* by the formulae

$$e = \phi(y, T) - T\frac{\partial \phi}{\partial T}(y, T) - y\frac{\partial \phi}{\partial y}(y, T); \quad \text{and} \quad q = -\frac{\partial \phi}{\partial y}(y, T).$$

Thus, there is only one macrostate (e, q), *and hence only one microstate* $\bar{\rho}(e, q)$, *corresponding to* (y, T) *if the Gibbs potential is differentiable there.*

We shall now prove the converse, i.e. that there is a macroscopic degeneracy at any point (y, T) where the Gibbs potential has a discontinuous first derivative. For definiteness we shall base our proof on the case where $\partial\phi/\partial T$ is discontinuous at (y, T): the corresponding proof for cases where one of the derivatives $\partial\phi/\partial y_j$ is discontinuous at (y, T) can be carried out analogously.

Assuming, then, that $\partial\phi/\partial T$ is discontinuous at (y, T), we aim to prove that there is a macroscopic degeneracy there by establishing that the system has equilibrium macrostates $(e^{(+)}, q^{(+)})$ and $(e^{(-)}, q^{(-)})$ with different entropy densities, namely,

$$s(e^{(\pm)}, q^{(\pm)}) = -\frac{\partial^{(\pm)}}{\partial T}\phi(y, T). \tag{20}$$

To this end, we start by noting that, as $\phi(y, T)$ is concave in T, for fixed y, it follows from the above property (C.3) of concave functions that there are sequences $\{T_n^{(\pm)}\}$ tending to T from above and below, such that ϕ is differentiable with respect to T at $(y, T_n^{(\pm)})$ and

$$\frac{\partial^{(\pm)}}{\partial T}\phi(y, T) = \lim_{n\to\infty}\frac{\partial \phi}{\partial T}(y, T_n^{(\pm)}). \tag{21}$$

Further, if $(e_n^{(\pm)}, q_n^{(\pm)})$ are equilibrium macrostates corresponding to $(y, T_n^{(\pm)})$, i.e. if

$$\phi(y, T_n^{(\pm)}) = e_n^{(\pm)} - T_n^{(\pm)}s(e_n^{(\pm)}, q_n^{(\pm)}) - y \cdot q_n^{(\pm)}, \tag{22}$$

then it follows from the differentiability of ϕ with respect to T at $(y, T_n^{(\pm)})$ that eqn (17) is applicable, so that

$$\frac{\partial\phi}{\partial T}(y, T_n^{(\pm)}) = -s(e_n^{(\pm)}, q_n^{(\pm)}),$$

and therefore, by (21),

$$\lim_{n\to\infty} s(e_n^{(\pm)}, q_n^{(\pm)}) = -\frac{\partial^{(\pm)}}{\partial T}\phi(y, T). \tag{23}$$

We now recall from §4.2 that the entropy function $s(e, q)$ is upper semicontinuous and that its domain of definition, D, is bounded and closed. In view of these latter properties of D, we see from the Bolzano–Weierstrass theorem that the sequences $\{(e_n^{(\pm)}, q_n^{(\pm)})\}$ have convergent subsequences whose limits also lie in D. We may therefore assume that $(e_n^{(\pm)}, q_n^{(\pm)})$ converge to limits $(e^{(\pm)}, q^{(\pm)})$, lying in D, as $n \to \infty$, since this can always be arranged by passing to suitable subsequences. Hence, by (23) and the upper semicontinuity of $s(e, q)$,

$$s(e^{(\pm)}, q^{(\pm)}) \geq -\frac{\partial^{(\pm)}}{\partial T}\phi(y, T) \tag{24}$$

Likewise, in view of the continuity of $\phi(y, T)$, the limiting form of the formula (22), as $n \to \infty$, is given by

$$\phi(y, T) \leq e^{(\pm)} - Ts(e^{(\pm)}, q^{(\pm)}) - y \cdot q^{(\pm)},$$

the equality sign being applicable here if and only if it is also valid in (24). Further, as $\phi(y, T)$ is the minimum value of $e - Ts(e, q) - y \cdot q$, the equality sign is operative here, in the last formula, and therefore also in (24). Thus, $(e^{(+)}, q^{(+)})$ and $(e^{(-)}, q^{(-)})$ are equilibrium macrostates corresponding to (y, T), and their entropy densities are $(\partial^{(\pm)}/\partial T)\phi(y, T)$, respectively, as in (20). Therefore, as $\partial\phi/\partial T$ is assumed to be discontinuous at (y, T), it follows that these equilibrium macrostates are different from one another, which means that there is a macroscopic degeneracy at (y, T). Similarly, one may show that a discontinuity in $\partial\phi/\partial y$ at (y, T) implies a macroscopic degeneracy at that point. Hence, as we have already proved that Σ has a unique equilibrium state at any point (y, T) where the derivatives of ϕ are continuous, we conclude that *the system exhibits a macroscopic degeneracy at (y, T) if and only if some first derivative of the Gibbs potential is discontinuous there*.

Assumption concerning location of discontinuities. We now supplement this result with the classical thermodynamical assumption that the region in (y, T)-space where the Gibbs potential has discontinuous derivatives consists of a finite number of surfaces, as in Fig. 5. Here we remark that, while this assumption is supported by the available rigorous results on tractable models,† as well as by the experimental evidence, it cannot be derived from the above general arguments, based as they are on concavity and continuity properties of the thermodynamic potentials. The concavity of $\phi(y, T)$ does not, of itself, guarantee that the points of discontinuity of its first derivatives form surfaces in (y, T)-space.‡

4.6. Phase structure

The results and assumptions of the previous Sections provide a picture of the equilibrium states and phases of Σ, which may be summarized as follows.

(1) The Gibbs potential $\phi(y, T)$ is continuous, but its derivatives may have discontinuities.

(2) The region of (y, T)-space where the first derivatives of ϕ are discontinuous consists of a set of surfaces, Γ say. Thus, by eqns (13) and (19), the variables (e, q) change *abruptly* as the phase point (y, T) crosses Γ, and nowhere else.

(3) The surfaces Γ comprise the region of macroscopic degeneracy, or coexisting phases, and the rest of (y, T)-space consists of single-phase regions.

Thus, the (y, T)-space is divided into single-phase regions separated by the surfaces Γ where the thermodynamical variables change discontinuously and where there is macroscopic degeneracy (cf. Fig. 7). The surfaces Γ therefore comprise the region of phase transitions. To be precise, these are *first-order* phase transitions, in Ehrenfest's sense, as they correspond to discontinuities in the first derivatives of the Gibbs potential. The fact that these transitions carry macroscopic degeneracy emerges only when Ehrenfest's thermodynamical scheme is supplemented by statistical mechanical arguments.

The other type of phase transitions, termed by Landau§ transitions of the *second order or second kind*, correspond to higher-order singularities in the Gibbs potential. Thus, they carry no discontinuities in the first derivatives of ϕ, i.e. in the values of (e, q), and no macroscopic degeneracy. According to Classical Thermodynamics,¶ transitions of this type occur at the so-called

† For example, in the case of the two-dimensional Ising model, y is the external, one-dimensional, magnetic field, and the region of discontinuity is the segment $0 \leq T < T_c$ of the line $y = 0$, where T_c is the transition temperature (cf. Lebowitz and Martin-Löf 1972). Similar results have been obtained in other tractable models (cf. Sinai 1982, Ch. 2).

‡ See Roberts and Varberg (1973), p. 7.

§ See Landau and Lifshitz (1959), Ch. 14. Note that the second-order transitions in Landau's scheme comprise all those of order higher than the first in Ehrenfest's classification.

¶ See Tisza (1966) and Landau and Lifshitz (1959).

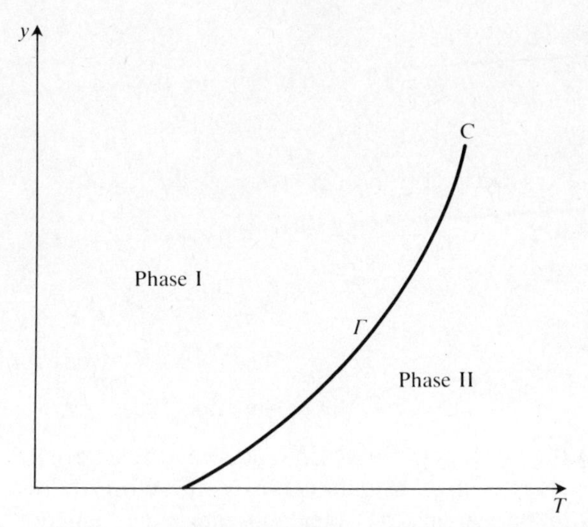

Fig. 7. Phase diagram with Γ the two-phase region and C the critical point.

critical points (y_c, T_c), which mark the termination of boundaries that separate different phases (cf. the boundary between II and III in Fig. 7); and their statistical mechanical feasibility has been proved, first by Onsager's exact solution of the two-dimensional Ising model and subsequently by solutions of various other models.† Here we make the general observation that, as the critical points occur at the margins of the one- and two- (or multi-) phase regions, the continuity of the first derivatives of the Gibbs potential survives from the former regions while some vestigial singularity remains from the latter ones.

4.7. Thermodynamical relations

In the scheme we have just described, the (y, T)-space consists of single-phase regions, separated by boundaries. We shall now derive both the standard thermodynamical formula relating changes in energy and entropy to mechanical work, in a single phase, and the Clausius–Clapeyron equation for coexisting phases, from formulae obtained above on the basis of convexity and continuity arguments.

We start by observing that, since the formula (19) is applicable in a single-phase region, it follows that.

$$d\phi = -s\,dT - q\,.\,dy \tag{25}$$

† See for example Lieb (1967*a*, *b*) and Baxter (1972) for solutions of certain lattice models with phase transitions of the second kind; and also Emch and Knops (1970) for models with translations of the Van der Waals–Maxwell type.

there. Hence, as

$$\phi \equiv e - Ts - y \,.\, q, \quad \text{by (13)},$$

and therefore

$$d\phi = de - T\,ds - s\,dT - y \,.\, dq - q \,.\, dy,$$

it follows that

$$T\,ds = de + dw \tag{26}$$

where

$$dw = -y \,.\, dq, \tag{27}$$

which is just the work per unit volume done by the external forces, y, in an infinitesimal change in q. Equation (26) is therefore the classical thermodynamical formula relating increments in energy and entropy to mechanical work, in single-phase regions.

In order to obtain the Clausius–Clapeyron equation for coexisting phases, we consider the situation where Γ is the surface in (y, T)-space separating two single-phase regions, I and II, say. Then if ϕ_{I} and ϕ_{II} are the respective Gibbs potentials for these phases, it follows from the continuity of the Gibbs potential for Σ that

$$\phi_{\mathrm{I}}(y, T) = \phi_{\mathrm{II}}(y, T) \quad \text{on } \Gamma. \tag{28}$$

Further, as ϕ_{I} and ϕ_{II} satisfy the versions of the formula (25) appropriate to the regions I and II, respectively,

$$d(\phi_{\mathrm{I}} - \phi_{\mathrm{II}}) = -(s_{\mathrm{I}} - s_{\mathrm{II}})\,dT - (q_I - q_{II}) \,.\, dy \quad \text{on } \Gamma,$$

and therefore, by (28),

$$(s_{\mathrm{II}} - s_{\mathrm{I}}) = (q_{\mathrm{I}} - q_{\mathrm{II}}) \,.\, dy/dT \text{ on } \Gamma.$$

Consequently, as $T(s_{\mathrm{II}} - s_{\mathrm{I}})$ is the *latent heat, λ,* for a transition from I to II, it follows that

$$\lambda = T(q_{\mathrm{I}} - q_{\mathrm{II}}) \cdot \frac{dy}{dT} \tag{29}$$

which is the Clausius–Clapeyron equation.

4.8. Concluding remarks

The extensive conserved quantities, introduced in this chapter, provide an essential link between the microscopic and macroscopic descriptions of equilibrium states of infinite systems. By incorporating these quantities into the model, we have established the following results, which provide the framework for a systematic theory of phase transitions.

(1) The extremal equilibrium states correspond *to pure phases*, both in the statistical mechanical sense that they cannot be resolved into mixtures of other equilibrium states, and in the thermodynamic sense that they carry no dispersions in the values of the global intensive thermodynamical variables (e, q).

(2) The discontinuities in the first derivatives of the Gibbs potential $\phi(y, T)$, or equivalently in the thermodynamical variables (e, q), occur at precisely those values of (y, T) for which the system supports different coexisting phases. In other words, the first-order phase transitions are characterized both by thermodynamical discontinuities and by quantum statistical degeneracies at the macroscopic level.

(3) The phase transitions of the second kind correspond to higher-order singularities in the Gibbs potential; and, as in Landau's theory, these carry neither discontinuities in the thermodynamic variables (e, q) nor macroscopic degeneracies.

Appendix A
Mathematical properties of the thermodynamic potentials

We shall devote this Appendix to the proof of the following properties of the entropy function $s(e, q)$, the Helmholtz potential $\psi(q, T)$ and the Gibbs potential $\phi(y, T)$, which were stated in §§4.2 and 4.3.

(1) The domain of definition, D, of the function $s(e, q)$ is convex and closed.
(2) The function $s(e, q)$ is concave and upper semicontinuous.
(3) The function $\psi(q, T)$ is convex in q and concave in T.
(4) The function $\phi(y, T)$ is concave.

The proofs of these properties run as follows.

Convexity of D. As noted at the end of §4.2, the domain D consists of the points (e, q) in $(n+1)$-dimensional Euclidean space, that lie within the range of the functions $(\hat{e}(\rho), \hat{q}(\rho))$, as ρ runs through the translationally invariant states of Σ. Thus, D consists of the points (e, q) that correspond to the translationally invariant states ρ according to the formula

$$\hat{e}(\rho) = e, \quad \text{and} \quad \hat{q}(\rho) = q. \tag{A.1}$$

Hence, if (e_1, q_1) and (e_2, q_2) belong to D, then there are translationally invariant states ρ_1, ρ_2 such that

$$\hat{e}(\rho_j) = e_j \quad \text{and} \quad \hat{q}(\rho_j) = q_j \quad \text{for} \quad j = 1, 2$$

Consequently, if $\bar{\rho} = \lambda\rho_1 + (1-\lambda)\rho_2$, with $0 < \lambda < 1$, then, as $\hat{e}$ and $\hat{q}$ are affine, it follows that

$$\hat{e}(\bar{\rho}) = \lambda e_1 + (1-\lambda)e_2 \quad \text{and} \quad \hat{q}(\bar{\rho}) = \lambda q_1 + (1-\lambda)q_2.$$

Thus, the point $(\lambda e_1 + (1-\lambda)e_2, \lambda q_1 + (1-\lambda)q_2)$ lies within the range, D, of the functionals $(\hat{e}(\rho), \hat{q}(\rho))$. In other words, if (e_1, q_1) and (e_2, q_2) lie in D, then so too does $\lambda(e_1, q_1) + (1-\lambda)(e_2, q_2)$ for $0 < \lambda < 1$, which means that D is convex.

Closure of D. Suppose that $\{(e_n, q_n)\}$ is a convergent sequence of points in D, and that (e, q) is the limit of this sequence. Since (e_n, q_n) lies in D it follows that there is some translationally invariant state ρ_n such that

$$\hat{e}(\rho_n) = e_n \quad \text{and} \quad \hat{q}(\rho_n) = q_n. \tag{A.2}$$

In view of the compactness of the states of Σ (property (P) of §3.3), we may take it that the sequence $\{\rho_n\}$ converges to a limit ρ, since this can always be arranged by passing to a subsequence, if necessary. Hence it follows from (A.2) that, as the functionals $\hat{e}$ and $\hat{q}$ are continuous,

$$(e_n, q_n) \to (\hat{e}(\rho), \hat{q}(\rho)) \quad \text{as} \quad n \to \infty.$$

Thus, the limit (e, q) of the sequence $\{(e_n, q_n)\}$ is given by $(\hat{e}(\rho), \hat{q}(\rho))$ and therefore lies in D. This means that D contains the limits of its convergent sequences, and is therefore closed.

Concavity of $s(e, q)$. We recall that $s(e, q)$ is the maximum value of the functional $\hat{s}(\rho)$ subject to the constraints that $\hat{e}(\rho)$, $\hat{q}(\rho)$ take the values e, q, respectively. Thus, if (e_1, q_2) and (e_2, q_2) are two points in the domain, D, of s, then

$$s(e_j, q_j) = \hat{s}(\rho_j) \quad \text{for} \quad j = 1, 2, \tag{A.3}$$

where ρ_j maximizes the functional $\hat{s}$ subject to the constraints that the functionals $(\hat{e}_j, \hat{q}_j)$ take the values (e_j, q_j). Since $\hat{s}$ is affine it follows from (A.3) that

$$\lambda s(e_1, q_1) + (1-\lambda)s(e_2, q_2) = \hat{s}(\lambda\rho_1 + (1-\lambda)\rho_2), \quad \text{for } 0 < \lambda < 1, \tag{A.4}$$

and, as $\hat{e}$ and $\hat{q}$ are also affine,

$$\lambda e_1 + (1-\lambda)e_2 \equiv \lambda\hat{e}(\rho_1) + (1-\lambda)\hat{e}(\rho_2) = \hat{e}(\lambda\rho_1 + (1-\lambda)\rho_2),$$

and

$$\lambda q_1 + (1-\lambda)q_2 \equiv \lambda\hat{q}(\rho_1) + (1-\lambda)\hat{q}(\rho_2) = \hat{q}(\lambda\rho_1 + (1-\lambda)\rho_2).$$

Further, as $s(\lambda e_1 + (1-\lambda)e_2, \lambda q_1 + (1-\lambda)q_2)$ is the maximum value of the functional $\hat{s}(\rho)$, for states for which $\hat{e}(\rho)$ and $\hat{q}(\rho)$ take the values $\lambda e_1 + (1-\lambda)e_2$ and $\lambda q_1 + (1-\lambda)q_2$, respectively, it follows from the last two equations that

$$s(\lambda e_1 + (1-\lambda)e_2, \lambda q_1 + (1-\lambda)q_2) \geqslant \hat{s}(\lambda\rho_1 + (1-\lambda)\rho_2).$$

Consequently, by (A.4),

$$s(\lambda e_1 + (1-\lambda)e_2, \lambda q_1 + (1-\lambda)q_2) \geqslant \lambda s(e_1, q_1) + (1-\lambda)s(e_2, q_2).$$

which means that s is concave.

Upper semicontinuity of $s(e, g)$. Suppose that $\{(e_n, q_n)\}$ is a convergent sequence of points in D, and that ρ_n is the translationally invariant state that maximizes the functional $\hat{s}$ subject to the constraint that $\hat{e}$ and $\hat{q}$ take the values e_n and q_n, respectively. Then

$$s(e_n, q_n) = \hat{s}(\rho_n);\ e_n = \hat{e}(\rho_n);\ \text{and}\ q_n = \hat{q}(\rho_n). \tag{A.5}$$

Again, we may take it that the sequence $\{\rho_n\}$ converges to a limit, ρ, since this can always be arranged by passing to a suitable subsequence, if necessary. Hence, in view of the continuity of $\hat{e}$ and $\hat{q}$, and the upper semicontinuity of $\hat{s}$, it follows from (A.5) that, if e_n, q_n, and $s(e_n, q_n)$ converge to e, q, and $\bar{s}$, respectively, as $n \to \infty$, then

$$\hat{s}(\rho) \geqslant \bar{s};\ \hat{e}(\rho) = e;\ \text{and}\ \hat{q}(\rho) = q. \tag{A.6}$$

From the last two of these equations and the definition of $s(e, q)$ it follows that $s(e, q) \geqslant \hat{s}(\rho)$; and therefore, by the first equation in (A.6), $s(e, q) \geqslant \bar{s}$, which means that the function s is upper semicontinuous.

Convexity of $\psi(q, T)$ *in* q. We recall that

$$\psi(q, T) = \min(e - Ts(e, q)). \tag{A.7}$$

Hence, if for fixed T and for values q_1, q_2 of q, the function $e - Ts(e, q_1)$ and $e - Ts(e, q_2)$ are minimized when $e = e_1$ and e_2, respectively, then

$$\psi(q_j, T) = e_j - Ts(e_j, q_j) \quad \text{for} \quad j = 1, 2,$$

and therefore

$$\lambda\psi(q_1, T) + (1-\lambda)\psi(q_2, T) = \lambda e_1 + (1-\lambda)e_2 - T(\lambda s(e_1, q_1) + (1-\lambda)s(e_2, q_2)), \text{ for } 0 < \lambda < 1.$$

Consequently, as $s(e, q)$ is concave,

$$\lambda\psi(q_1, T) + (1-\lambda)\psi(q_2, T) \geqslant \bar{e} - Ts(\bar{e}, \lambda q_1 + (1-\lambda)q_2) \tag{A.8}$$

where

$$\bar{e} = \lambda e_1 + (1-\lambda)e_2.$$

Furthermore, it follows from (A.7) that the RHS of (A.8) cannot be less than $\psi(\lambda q_1 + (1-\lambda)q_2, T)$, and therefore the latter inequality implies that

$$\lambda\psi(q_1, T) + (1-\lambda)\psi(q_2, T) \geqslant \psi(\lambda q_1 + (1-\lambda)q_2, T),$$

i.e. that $\psi(q, T)$ is convex in q.

Concavity of $\psi(q, T)$ in T. Let $\bar{e}$ be a value of e that maximizes $e - (\lambda T_1 + (1-\lambda)T_2)s(e, q)$ for given q, T_1, T_2, and λ $(\in(0, 1))$. Then, by (A.7),

$$\psi(q, \lambda T_1 + (1-\lambda)T_2) = \bar{e} - (\lambda T_1 + (1-\lambda)T_2)s(\bar{e}, q),$$

and

$$\psi(q, T_j) \leqslant \bar{e} - T_j s(\bar{e}, q) \quad \text{for} \quad j = 1, 2.$$

From these last two formulae it follows that

$$\psi(q, \lambda T_1 + (1-\lambda)T_2) \geqslant \lambda\psi(q, T_1) + (1-\lambda)\psi(q, T_2).$$

i.e. that ψ is concave in T.

Concavity of $\phi(y, T)$. We recall that

$$\phi(y, T) = \min_{e,q}(e - Ts(e, q) - y \,.\, q), \tag{A.9}$$

and use the method we have just employed to prove that ψ is concave in T. Thus, we suppose that

$$y = \lambda y_1 + (1-\lambda)y_2, \qquad T = \lambda T_1 + (1-\lambda)T_2, \tag{A.10}$$

where $0 < \lambda < 1$, and that $e - Ts(e, q) - y \,.\, q$ is minimized when $e = \bar{e}$ and $q = \bar{q}$, say. Then, by (A.9),

$$\phi(y, T) = \bar{e} - Ts(\bar{e}, \bar{q}) - y \,.\, \bar{q},$$

and

$$\phi(y_j, T_j) \leqslant \bar{e} - T_j s(\bar{e}, \bar{q}) - y_j \,.\, \bar{q} \quad \text{for} \quad j = 1, 2.$$

It follows immediately from these last two formulae and (A.10) that

$$\phi(\lambda y_1 + (1-\lambda)y_2, \lambda T_1 + (1-\lambda)T_2) \geqslant \lambda\phi(y_1, T_1) + (1-\lambda)\phi(y_2, T_2),$$

which means that ϕ is concave.

Appendix B
The dispersion-free character of global intensive thermodynamical variables in extremal equilibrium states

We shall prove here that the global densities of the extensive conserved quantities, Q, are dispersionless in extremal equilibrium states, ρ, i.e. that

$$\lim_{\Lambda\uparrow} |\Lambda|^{-2}[\rho(Q(\Lambda)^2) - (\rho(Q(\Lambda))^2] = 0. \tag{B.1}$$

We shall obtain this result by showing that, for sufficiently large Λ, $Q(\Lambda)/|\Lambda|$ may be approximated arbitrarily closely by the space average over that region of a certain local observable $\mathcal{Q}$; and hence that the global density of Q, like the global space average† of $\mathcal{Q}$, is dispersionless in extremal equilibrium states. For simplicity, we shall confine the proof of (B.1) to the case where Λ increases to infinity over a sequence $\{\Lambda_n\}$ of cubes whose side lengths are $\{2^n\}$, respectively, in units of the lattice spacing. Thus, the result we shall obtain is that

$$\lim_{n\to\infty} |\Lambda_n|^{-2}[\rho(Q(\Lambda_n)^2 - (\rho(Q(\Lambda_n))^2] = 0. \tag{B.2}$$

The generalization of this result to other sequences of suitably regular regions, $\{\Lambda\}$ can be derived from (B.2) by approximating the Λ's arbitrarily closely by unions of Λ_n's.

Our proof of (B.2) is based on the following construction.

We introduce a sequence of cubes $\{\Lambda_N^{(L)}\}$ with side lengths $\{L\,.\,2^N\}$, respectively, L being an integer. These cubes are therefore related to the cubes $\{\Lambda_n\}$, specified above by the formula.

$$\Lambda_N^{(L)} = \Lambda_n, \quad \text{with} \quad L = 2^m \text{ and } n = N + m.$$

Hence, in order to prove (B.2), it suffices to show that

$$\lim_{L\to\infty} \lim_{N\to\infty} |\Lambda_N^{(L)}|^{-2}[\rho(Q(\Lambda_N^{(L)})^2) - (\rho(Q(\Lambda_N^{(L)}))^2] = 0. \tag{B.3}$$

With the proof of this formula as our target, we now note that it follows from our definition of $\Lambda_N^{(L)}$ that this cube is the union of 2^{Nd} copies of a cube $\Lambda^{(L)}$, of side L, d being the dimensionality of the lattice. We take the centre of $\Lambda^{(L)}$ to be the origin and denote the centres of the copies $\{\Lambda_J^{(L)}\}$ of this cube, that comprise $\Lambda_N^{(L)}$, by $\{x_J\}$. We define

$$\mathcal{Q}^{(L)} = Q(\Lambda^{(L)})/|\Lambda^{(L)}|, \tag{B.4}$$

$$\mathcal{Q}_N^{(L)} = \sum_{x\in\Lambda_N^{(L)}} \mathcal{Q}(x); \tag{B.5}$$

and, for an arbitrary region Λ, we define $Q'(\Lambda)$ to be the space average of $Q(\Lambda)$ over $\Lambda^{(L)}$, i.e.

$$Q'(\Lambda) = |\Lambda^{(L)}|^{-1} \sum_{x\in\Lambda^{(L)}} Q(\Lambda + x) \tag{B.6}$$

We now aim to show that, by taking L sufficiently large, we may approximate the density of the extensive quantity Q arbitrarily closely by the space average of $\mathcal{Q}^{(L)}$, in the sense that

$$\lim_{L\to\infty} \lim_{N\to\infty} |\Lambda_N^{(L)}|^{-1}\, \|Q(\Lambda_N^{(L)}) - \mathcal{Q}_N^{(L)}\| = 0. \tag{B.7}$$

† Recall that, by (15), space averages of local observables are dispersion-free in extremal equilibrium states.

In fact, this formula is the key to the proof of (B.3). For since, by (15), the space average of any local observable is dispersion-free in an extremal equilibrium state ρ, it follows from (B.5) that

$$\lim_{N\to\infty} |\Lambda_N^{(L)}|^{-2}[\rho((\mathcal{Q}_N^{(L)})^2) - (\rho(\mathcal{Q}_N^{(L)}))^2] = 0, \tag{B.8}$$

and, in view of the uniform boundedness† of the density of Q, (B.3) is a simple consequence of (B.7) and (B.8). Hence, it suffices for us to prove (B.7). We shall break up the proof into the following three lemmas.

***Lemma* 1.**

$$|\Lambda_N^{(L)}|^{-1} \, \|Q(\Lambda_N^{(L)}) - |\Lambda^{(L)}| \sum_J \mathcal{Q}(x_J)\| \to 0,$$
$$\text{uniformly in } N, \text{ as } L \to \infty, \tag{B.9}$$

where, we recall, $\{\Lambda_J^{(L)}\}$ is the set of copies of $\Lambda^{(L)}$ comprising $\Lambda_N^{(L)}$, and their respective centres are $\{x_J\}$.

***Lemma* 2.**

$$|\Lambda_N^{(L)}|^{-1} \, \|Q'(\Lambda_N^{(L)}) - \mathcal{Q}_N^{(L)}\| \to 0,$$
$$\text{uniformly in } N, \text{ as } L \to \infty \tag{B.10}$$

***Lemma* 3.**

$$|\Lambda_N^{(L)}|^{-1} \, \|Q(\Lambda_N^{(L)}) - Q'(\Lambda_N^{(L)})\| \to 0, \quad \text{as } N \to \infty. \tag{B.11}$$

One sees immediately that Lemmas 2 and 3 imply the required result (B.7). Lemma 1, on the other hand, is used here just as a stepping-stone towards Lemma 2.

It remains for us, then, to prove these three lemmas.

***Proof of Lemma* 1.** By our definition of $\Lambda_N^{(L)}$, this cube is the union of 2^d disjoint copies $\{\Lambda_{N-1,r}^{(L)} \mid r = 1, \ldots, 2^d\}$ of $\Lambda_{N-1}^{(L)}$. Hence, as the extensive variable Q satisfies the surface condition (Q3) of §4.2,

$$\|Q(\Lambda_N^{(L)}) - \sum_{r=1}^{2^d} Q(\Lambda_{N-1,r}^{(L)})\| < C \, |\Lambda_N^{(L)}|^{1-\alpha}. \tag{B.12}$$

where C and α are constants, with $0 < \alpha < 1$. Similarly, as $\Lambda_{N-1,r}^{(L)}$ is the union of 2^d disjoint copies $\{\Lambda_{N,r,s}^{(L)} \mid s = 1, \ldots, 2^d\}$ of $\Lambda_{N-2}^{(L)}$,

$$\|Q(\Lambda_{N-1,r}^{(L)}) - \sum_{s=1}^{2^d} Q(\Lambda_{N-2,r,s}^{(L)})\| < C \, |\Lambda_{N-1}^{(L)}|^{1-\alpha},$$

and therefore, by (B.12),

$$\|Q(\Lambda_N^{(L)}) - \sum_{r=1}^{2^d} \sum_{s=1}^{2^d} Q(\Lambda_{N-2,r,s}^{(L)})\| < C \, |\Lambda_N^{(L)}|^{1-\alpha} + 2^d C \, |\Lambda_{N-1}^{(L)}|^{1-\alpha};$$

† See condition (Q4) of §4.2.

or equivalently, denoting the cubes $\{\Lambda_{N-2,r,s}^{(L)}\}$, whose union comprises $\Lambda_N^{(L)}$, by $\{\Lambda_{N-2,t}^{(L)} \mid t=1,\ldots,2^{2d}\}$,

$$\|Q(\Lambda_N^{(L)}) - \sum_{t=1}^{2^{2d}} Q(\Lambda_{N-2,t}^{(L)})\| < C\,|\Lambda_N^{(L)}|^{1-\alpha} + 2^d C\,|\Lambda_{N-1}^{(L)}|^{1-\alpha}.$$

Continuing this procedure of dividing $\Lambda_N^{(L)}$ into copies of $\Lambda_{N-1}^{(L)}$, then of $\Lambda_{N-2}^{(L)}$, etc., we eventually obtain a resolution of $\Lambda_N^{(L)}$ into the copies $\{\Lambda_J^{(L)} \mid J=1,\ldots,2^{Nd}\}$ of $\Lambda^{(L)}$, together with the associated inequality

$$\|Q(\Lambda_N^{(L)}) - \sum_{J=1}^{2^{Nd}} Q(\Lambda_J^{(L)})\| < C[|\Lambda_N^{(L)}|^{1-\alpha} + 2^d\,|\Lambda_{N-1}^{(L)}|^{1-\alpha} + \ldots + 2^{(N-1)d}\,|\Lambda_1^{(L)}|^{1-\alpha}].$$

Hence, as $|\Lambda_N^{(L)}| = L^d \,.\, 2^{Nd}$,

$$\begin{aligned}|\Lambda_N^{(L)}|^{-1}\,\|Q(\Lambda_N^{(L)}) - \sum_J Q(\Lambda_J^{(L)})\| &< C\,.\,L^{-\alpha d}\,.\,2^{-N\alpha d}[1 + 2^{\alpha d} + \ldots + 2^{(N-1)\alpha d}] \\ &= C\,.\,L^{-\alpha d}(1-2^{-N\alpha d})/(2^{\alpha d}-1) \\ &< CL^{-\alpha d}/(2^{\alpha d}-1),\end{aligned}$$

from which it follows that

$$|\Lambda_N^{(L)}|^{-1}\,\|Q(\Lambda_N^{(L)}) - \sum_J Q(\Lambda_J^{(L)})\| \to 0, \quad \text{uniformly w.r.t. } N, \text{ as } L \to \infty.$$

This formula is in fact equivalent to the required result, (B.9), since, by (B.4), $Q(\Lambda_J^{(L)})$ is $|\Lambda^{(L)}|$ times the space translate, $\mathcal{Q}^{(L)}(x_J)$, of $\mathcal{Q}^{(L)}$ to the centre of $\Lambda_J^{(L)}$.

Q.E.D.

***Proof of Lemma* 2.** Since the norm of a local observable is invariant under space translations, i.e. $\|A(x)\| \equiv \|A\|$, it follows that the norm in (B.9) is unchanged by the transformation $Q(\Lambda_N^{(L)}) \to Q(\Lambda_N^{(L)}+x)$, $\mathcal{Q}^{(L)}(x_J) \to \mathcal{Q}^{(L)}(x_J+x)$, and therefore that $|\Lambda_N^{(L)}|^{-1}\,\|Q(\Lambda_N^{(L)}+x) - |\Lambda^{(L)}|\sum_J \mathcal{Q}^{(L)}(x_J+x)\| \to 0$, uniformly in N and x, as $L \to \infty$. Hence, on averaging this formula over space translations, x, in $\Lambda^{(L)}$, and using the equation (B.6) for $Q'(\Lambda)$, we obtain the required result (B.12).

Q.E.D.

***Proof of Lemma* 3.** Let $\tilde{\Lambda}_N^{(L)}$ be the cube obtained by removing from $\Lambda_N^{(L)}$ the 'shell' comprising the sites lying within the distance L of its surface. Thus, for any x in $\Lambda^{(L)}$, the cube $\Lambda_N^{(L)}+x$ contains $\tilde{\Lambda}_N^{(L)}$. Further, the volume and surface area of the region of $(\Lambda_N^{(L)}+x)$ lying outside $\tilde{\Lambda}_N^{(L)}$ are bounded by values $c_1 L\,|\Lambda_N^{(L)}|^{(d-1)/d}$ and $c_2\,|\Lambda_N^{(L)}|^{(d-1)/d}$, respectively, for all x in $\Lambda^{(L)}$, where c_1 and c_2 are finite constants. Therefore, it follows from the extensivity of Q and the uniform boundedness of its density, as specified by properties (Q3) and (Q4), respectively, of §4.2, that

$$\|Q(\Lambda_N^{(L)}+x)-Q(\tilde{\Lambda}_N^{(L)})\|<C\,|\Lambda_N^{(L)}|^{(d-1)/d}+DL\,|\Lambda_N^{(L)}|^{(d-1)/d},$$

for all x in $\Lambda^{(L)}$, where C and D are constants. Hence,

$$|\Lambda_N^{(L)}|^{-1}\,\|Q(\Lambda_N^{(L)}+x)-Q(\tilde{\Lambda}_N^{(L)})\|\to 0, \text{ uniformly w.r.t. } x \text{ in } \Lambda^{(L)}, \quad \text{as } N\to\infty. \tag{B.13}$$

On putting $x=0$ in this formula, we see that

$$|\Lambda_N^{(L)}|^{-1}\,\|Q(\Lambda_N^{(L)})-Q(\tilde{\Lambda}_N^{(L)})\|\to 0 \quad \text{as} \quad N\to\infty,$$

which implies that we may replace $Q(\tilde{\Lambda}_N^{(L)})$ by $Q(\Lambda_N^{(L)})$ in (B.13), i.e. that

$$|\Lambda_N^{(L)}|^{-1}\,\|Q(\Lambda_N^{(L)}+x)-Q(\Lambda_N^{(L)})\|\to 0, \quad \text{uniformly w.r.t. } x \text{ in } \Lambda^{(L)}, \quad \text{as } N\to\infty.$$

On averaging this equation over the sites, x, in $\Lambda^{(L)}$ and using the definition (B.6) of $Q'(\Lambda)$, we obtain the required result (B.11).

Q.E.D.

Part III

Collective phenomena

5

Phase transitions

5.1 Introduction

In Landau's scheme, a phase transition is classed as being of the first or second kind according to whether or not it carries a discontinuity in a first derivative of the Gibbs potential. Since, as we saw in the previous chapter, any such discontinuity corresponds to a macroscopic degeneracy, this means that the hallmark of a transition of the first kind is the existence of different phases at the transition point: by contrast, the difference between the phases involved in a transition of the second kind disappears at the critical point. It follows† that any transition carrying an *abrupt* structural change, e.g. from a face-centred to a body-centred cubic crystal, must be of the first kind, since the phases existing at the transition point are manifestly distinguishable from one another.

This implies that the changes in structure involved in transitions of the second kind must take place continuously. Landau‡ proposed, on the basis of considerations of concrete cases, that any transition of the second kind corresponds to the continuous variation of some intensive quantity, η, which in some sense represents the 'degree of ordering' of the microscopic components of the system, from non-zero values below the critical temperature to zero above it.

For example, in the case of a transition from a ferromagnetic to a paramagnetic phase, η is the spontaneous polarization, an obvious measure of the degree of alignment, or ordering, of the spins. In general, η is termed an *order parameter*, and may be formulated so as to represent a measure of asymmetry (polarization!) and long-range correlations, as in the ferromagnetic case (cf. §5.2).

Landau's theory of second-order phase transitions, then, is based on the hypothesis that each such transition corresponds to the continuous variation of an order parameter, η, from non-zero values below the critical temperature, T_c, to zero above it. This hypothesis provides a simple connection between the change of structure, or order, at T_c, and the thermodynamical singularity there. For the assumption that η varies continuously with T from non-zero values below to zero above T_c implies that η has a singularity at the critical point and that consequently the

† Cf. Landau and Lifshitz (1959), Ch. 8.
‡ See Landau and Lifshitz (1959), Ch. 14.

thermodynamic potentials are also singular there by virtue of their η-dependence. We remark here that it is feasible that there may be phase transitions of the first kind in which an order parameter varies continuously through non-zero values below a transition point, and then drops discontinuously to zero, since there are models† of magnetic systems whose spontaneous polarizations behave in this way. In general, a phase transition characterized by an order parameter that varies continuously over non-zero values for $T < T_c$ and vanishes for $T > T_c$, is termed an *order–disorder transition*.

The two most basic problems in the theory of order–disorder transitions are (a) the establishment of criteria specifying which classes of systems support these transitions, and (b) the determination of the general features of critical phenomena. The particular difficulties of the second problem stem from the extreme instability of critical states with respect to changes in thermodynamical conditions.

We shall devote the present chapter to an account of developments that have been made towards a resolution of these two basic problems. Let us now briefly discuss what is involved here, before going on to the mathematical treatment of these problems.

First we remark that it has been known for some time that various 'low-dimensional' systems with short-range interactions do *not* support order–disorder transitions. This has been proved for example, for one-dimensional systems‡ and for two-dimensional ones with continuous internal symmetries, such as the Heisenberg ferromagnet.§ The proofs are based on energy-versus-entropy arguments that show that these systems cannot support ordered states at finite temperature.

On the other hand, there are now proofs of the existence of order–disorder transitions in a variety of systems. Apart from the special cases of exactly solvable models,¶ the proofs have been achieved by utilizing rigorous versions of the classic Mayer (1947) cluster expansions to prove the absence of ordering at sufficiently high temperatures and by obtaining bounds on the relevant order parameters so as to show that these are non-zero when the temperature is low enough. Generally speaking, it is the proof of low temperature ordering that is the difficult problem. Until relatively recently, the only systematic procedure for proving such ordering was based on the Peierls argument, devised originally for the two-dimensional Ising ferromagnet and subsequently adapted to other, closely

† See, for example, Thouless (1969) and Kotecky and Shlosman (1982).

‡ See Landau and Lifshitz (1959), p. 482, and, for a rigorous version of the argument, Simon and Sokal (1981).

§ See Mermin and Wagner (1966).

¶ Notable examples (all two-dimensional) of exactly solved models with order–disorder transitions are the two-dimensional Ising ferromagnet (Onsager 1944) and certain more complex lattice systems (Lieb 1967*a*,*b*; Baxter 1972).

related models† with *discrete* internal symmetries (e.g. spin reversals). An important advance in the subject was achieved by J. Fröhlich, Simon, and Spencer (FSS) (1976) who devised a method for proving the existence of order–disorder transitions with *continuous* symmetry breakdown in a certain class of models. The argument of FSS and its subsequent generalization by Dyson, Lieb, and Simon (1978) established low-temperature ferromagnetic or antiferromagnetic ordering in various three-dimensional models of the Heisenberg type.‡ The method employed to obtain these results is essentially to show that the systems concerned undergo a *Bose–Einstein condensation of magnons* (*spin waves*). Furthermore, one essential idea of the FSS argument, connected with space reflection symmetry, was subsequently deployed by J. Fröhlich, Israel, Lieb, and Simon (1978, 1980) to produce a much more powerful version of the Peierls argument and thereby to prove the existence of phase transitions with discrete symmetry breakdown in a much larger class of systems than previously. Other notable improvements of Peierls's argument have been made by Pirogov and Sinai (1975, 1978). Thus, there are now some general techniques for establishing criteria for order–disorder transitions in a rather wide class of models.

As regards the theory of critical phenomena, the central problem there is to obtain the forms of the critical singularities, since these represent the sensitivity of a system to changes in thermodynamical conditions near the critical point. In view of the general connection between spontaneous fluctuations and response to external forces,§ these singularities may be shown to correspond to infinite range correlations in the observables constituting the order parameter – in the case of a ferromagnet, for example, these would be the spins. An important development in the theory of critical phenomena, due largely to Kadanoff (1966), Fisher (1967*b*), and Wilson (1971), is based on the idea that the infinite range of these correlations renders the bulk critical properties of a system scale invariant and thus independent of all but certain qualitative features of its microscopic constitution, such as symmetry, dimensionality, and interaction range (finite or infinite). On the basis of this idea, Wilson and Fisher (1972) devised a technique for obtaining the forms of critical singularities of many-particle systems. This was based on the renormalization group¶ of quantum field theory, which essentially represents scale transformations and

† Examples (in two or three dimensions) are Ising spin systems with ferromagnetic interactions involving non-nearest neighbours (Griffiths 1967) and the highly anisotropic version of the Heisenberg ferromagnet (Ginibre 1969; Robinson 1969).

‡ Unfortunately, these do not include the isotropic Heisenberg ferromagnet, which still presents problems (cf. §5.3.4). They do include the Heisenberg antiferromagnet, the XY-model and the classical Heisenberg ferromagnet.

§ cf. the discussion of the fluctuation–dissipation theorem in §3.1.

¶ The relevance of this group to critical phenomena appears to have been first recognized by Di Castro and Jona-Lasinio (1969).

serves to relate critical singularities to large-scale properties of macroscopic systems. Furthermore, the mathematical structure of the renormalization theory has been elucidated by Bleher and Sinai (1973, 1975), Jona-Lasinio (1975), and Sinai (1976), who have recast it in probabilistic terms and thereby characterized critical points by non-central limiting behaviour.† Thus, the renormalization theory has led to a new conception of critical phenomena as well as providing accurate computations of the properties of critical singularities. However, as pointed out by Griffiths (1981), there remains a problem concerning the precise form of the relationship it purports to establish between the microscopic and the large-scale critical properties of many-particle systems.

We shall present a treatment of the issues discussed in this Introduction as follows. In §5.2, we shall formulate a characterization of order parameters in terms of broken symmetries and long-range correlations. In §5.3, we shall review the methods that have been devised to prove the existence or non-existence of order–disorder transitions in certain classes of models. In §5.4, we shall present an account of developments in the theory of critical singularities, starting with phenomenological ideas leading up to the renormalization method, then passing to the probabilistic formulation of it, with application to a class of tractable models, and finally indicating some outstanding problems. We shall conclude this chapter in §5.5 with a brief discussion of the picture provided by the results described here and of some basic open problems. The two Appendices will be devoted to proofs of technical points.

5.2. Order parameters, symmetry breakdown, and long-range correlations

In order to formulate the concept of an order parameter, we start by considering the specific case of the two-dimensional Ising model. This model undergoes a phase transition of the second kind at a critical temperature T_c such that, for $T < T_c$, it has two pure equilibrium phases, with equal and opposite polarizations, while for $T > T_c$ it has just one equilibrium phase, which is unpolarized. The fact that the pure phases are polarized, for $T < T_c$, means that the spins are preferentially aligned then, and we refer to the system as being *ordered* by virtue of this alignment. Since the polarization represents a measure of the degree of alignment, we refer to it as the *order parameter*, η. This parameter evidently represents the strength of an 'inner field' (polarization) that acts on its own source, namely the spins, in such a way as to preserve their alignment and thereby to stabilize the ordering.

The concept of an order parameter as a stabilized, symmetry-breaking

† We recall that the central limit theorem, when applicable, tells us essentially that the equilibrium fluctuations in a macroscopic variable has a Gaussian distribution with dispersion that increases as the square root of the appropriate extensivity parameter (cf. Khinchin 1949).

inner field may be formulated in a general context as follows. Suppose that Σ is an arbitrary system which undergoes a phase transition of the second kind, satisfying the following conditions.

(a) Σ has a space-dependent local observable, $\hat{\eta}(x)$, which transforms linearly under a symmetry group, $G = \{\alpha(g)\}$, of automorphisms of the algebra of the observables. Thus, if $\hat{\eta}(x)$ has n components $(\hat{\eta}_1(x), \ldots, \hat{\eta}_n(x))$, say

$$(\alpha(g)\hat{\eta}(x))_r = \sum_{s=1}^{n} g_{rs}\hat{\eta}_s(x).$$

It will be assumed that the group formed by the matrices $[g_{rs}]$ is non-trivial in the sense that the only single column vector that is invariant under the action of all these matrices is the zero one.

(b) The equilibrium value of $\hat{\eta}(x)$ is non-zero in the pure phases below the critical temperature, T_c. By (a), this implies a breakdown of G-symmetry below the critical point.

(c) There is no macroscopic degeneracy at temperatures above T_c. This implies that there is no symmetry breakdown and therefore, by (a), that the equilibrium expectation value of $\hat{\eta}(x)$ vanishes in this temperature range.

Under these conditions, the expectation value, η, of $\hat{\eta}(x)$ plays the role of a generalized polarization which represents a measure of the degree of asymmetry of the pure equilibrium phases below the critical point, T_c, and vanishes at temperatures above T_c. Accordingly, we designate η as the *order parameter* of the system, and we term the space-dependent observable $\hat{\eta}(x)$ the *order field*. Some simple examples of systems satisfying the conditions (a)–(c) are the following.

(1) The Ising ferromagnet. Here, $\hat{\eta}(x)$ is the spin $\sigma(x)$ at the site x and G consists of the identity and the spin reversal automorphism.

(2) The Ising antiferromagnet: this has interactions of the same form as those of the Ising ferromagnet, except that the sign of the coupling constant is reversed. In the antiferromagnetic phase $(T < T_c)$, the subsystems of spins on two interlocking and complementary sublattices, X_1 and X_2, are equally and oppositely polarized. The order field $\hat{\eta}(x)$, defined at the midpoints, x, of the bonds joining neighbouring sites, x_1 and x_2, of X_1 and X_2, respectively, is then $\sigma(x_1) - \sigma(x_2)$, and the order parameter is the difference between the polarizations of these sublattices. The symmetry group G again corresponds to spin reversals.

Note. This model may also be taken to represent a binary alloy (e.g. CuZn), the two different species of atoms playing the roles of 'up' and 'down' spins.

(3) The Heisenberg ferromagnet† and antiferromagnet. The order field, $\hat{\eta}(x)$, and the order parameter, η, for these are given by the same prescriptions as for the corresponding Ising systems, except that they are now vectors. The symmetry group G corresponds to rotations of the spins.

(4) Certain ferroelectrics whose crystal structures change continuously with temperature below the transition point. A good example is $BaTiO_3$.‡ For $T > T_c$, the Ba atoms of this materials form a simple cubic lattice, X, and the Ti and O atoms are placed at the body centres and the face centres, respectively, of its cells. For $T < T_c$, the Ti and O atoms are all displaced, in the pure phases, along one of the crystallographic directions, their displacements varying continuously with T and vanishing at T_c. The order field, $\hat{\eta}(x)$, defined at the centres, x, of the cells of X, is the set of atomic displacements in the respective cells. The order parameter, η, as given by the expectation value of $\hat{\eta}(x)$, thus determines the electric polarization. The symmetry group G consists of the discrete rotations between the crystallographic axes of X.

Ordering and long-range correlations. Suppose now that Σ undergoes an order–disorder transition in accordance with the above specifications (a) and (b). Then, if ρ is a pure equilibrium phase at a temperature T, below T_c, its expectation value, η, for the order field $\hat{\eta}(x)$ will be non-zero. Moreover, as ρ is a pure phase it carries no dispersion in the global space average of $\hat{\eta}(x)$, i.e.

$$\lim_{\Lambda\uparrow} |\Lambda|^{-2}\rho\left(\left(\sum_{x\in\Lambda} \hat{\eta}(x)\right)^2\right) \equiv \lim_{\Lambda\uparrow} |\Lambda|^{-2} \sum_{x,x'\in\Lambda} \rho(\hat{\eta}(x)\,.\,\hat{\eta}(x')) = \eta^2. \tag{1}$$

This implies that $\rho(\hat{\eta}(x)\,.\,\hat{\eta}(x'))$ cannot tend to zero as $|x-x'|\to\infty$; in fact, it follows from this formula and the translational invariance of ρ that η^2 is the only limit to which this correlation function can tend as $|x-x'|\to\infty$. The pure phases therefore carry long-range correlations in the order field at temperatures below T_c.

The same is evidently true for the other equilibrium states below T_c, as these are just mixtures of the pure phases. In particular, it is true for the equilibrium states $\bar{\rho}$ that are symmetric with respect to the group G: one can always guarantee the existence of one such equilibrium state at any temperature, T, since it may be constructed as the infinite volume limit of the space averaged canonical state of a finite version of Σ.§ Thus, if $\bar{\rho}$ is a symmetrical equilibrium state below the critical point, then, as $\bar{\rho}$ is a

† Here the citation of the Heisenberg ferromagnet as an example of a phase transition is provisional, since its low-temperature ordering has not yet been proved (cf. §5.3.4).

‡ See Landau and Lifshitz (1959), p. 430.

§ Cf. eqn (38) of §3.4.2.

mixture of pure phases at the relevant temperature, it follows from (1) that

$$\lim_{\Lambda\uparrow} |\Lambda|^{-2}\bar{\rho}\left(\left(\sum_{x\in\Lambda} \hat{\eta}(x)\right)^2\right) > 0, \tag{2}$$

which implies that $\bar{\rho}$ has long-range correlations in the order field.

Conversely, if one merely knows that $\bar{\rho}$ is a symmetric equilibrium state possessing the long-range correlation property (2), then one can infer that there is ordering, with G-symmetry breakdown, in the pure phases at the temperature concerned. The argument runs as follows. Since $\bar{\rho}$ is symmetrical with respect to G, $\bar{\rho}(\hat{\eta}(x)) = 0$, and consequently (2) implies that this state carries a dispersion in the global space average of $\hat{\eta}(x)$ and is therefore a mixture of pure phases $\{\rho\}$. The purity of these ensures that they carry no dispersion in the global space average of $\hat{\eta}(x)$. Therefore, if the expectation values of $\hat{\eta}(x)$ were zero for these phases, ρ, then $|\Lambda|^{-2}\rho((\Sigma_{x\in\Lambda} \hat{\eta}(x))^2)$ would tend to zero in the limit $\Lambda \to \infty$, and the same would be true for $|\Lambda|^{-2}\bar{\rho}((\Sigma_{x\in\Lambda} \hat{\eta}(x))^2)$, as $\bar{\rho}$ is a weighted sum of the ρ's. Since this would contradict (2), we conclude that this latter condition, as applied to a symmetrical equilibrium state $\bar{\rho}$, implies that at least some of the pure phases, at the temperature concerned, yield non-zero expectation values of $\hat{\eta}(x)$ and are therefore ordered. Hence, *the long-range correlation property* (2), *for symmetrical equilibrium states, is a necessary and sufficient condition for ordering*.

Comments. (1) This condition (2) may sometimes be usefully replaced by the limiting form of a corresponding inequality for a finite version of the model. For, suppose that $\Sigma(\Lambda)$ is the finite system consisting of particles of the given species in a cube Λ, with free or periodic boundary conditions, and that $\Sigma_{x\in\Lambda} \hat{\eta}(x)$ is a constant of the motion for this system. Then a theorem of Griffiths (1966), which we shall prove in Appendix A, tells us that, if the canonical equilibrium average $\langle \ldots \rangle_\Lambda$ for the $\Sigma(\Lambda)$ satisfies the condition

$$\lim_{\Lambda\uparrow} |\Lambda|^{-2}\left\langle\left(\sum_{x\in\Lambda} \hat{\eta}(x)\right)^2\right\rangle_\Lambda > 0, \tag{3}$$

then the *infinite* system Σ will exhibit the ordering we have just described.

(2) Another sufficient condition for this ordering may be expressed as a generalized Bose–Einstein condensation condition. For if $\hat{\eta}_{\mathrm{F}}(k)$ is the Fourier transform of $\hat{\eta}(x)$,

$$\hat{\eta}_{\mathrm{F}}(k) = \sum_x \hat{\eta}(x)\, \mathrm{e}^{-\mathrm{i}k\,.\,x}, \tag{4}$$

and if $\bar{\rho}$ exhibits a condensation of the Bose–Einstein kind in the zero wave-vector mode, i.e.

$$\rho(\hat{\eta}_{\mathrm{F}}(k)^*\,.\,\hat{\eta}_{\mathrm{F}}(k)) = c\delta(k) + F(k), \tag{5}$$

with $c > 0$ and $F(k)$ absolutely integrable, then it follows easily from these last two formulae that $\bar{\rho}(\eta(x)\,.\,\hat{\eta}(x'))\to c$ as $|x-x'|\to\infty$, which implies that the condition (2) is fulfilled.

(3) An extension of condition (2) leads to a characterization of a critical point by infinite-range correlations in the fluctuations of the order field. For, if Σ undergoes a phase transition of the second kind, then eqn (2) is applicable below the critical temperature, T_c, and hence

$$\lim_{\Lambda\uparrow} |\Lambda|^{-1} \sum_{x,x'\in\Lambda} \bar{\rho}(\hat{\eta}(x)\,.\,\hat{\eta}(x'))=\infty \quad \text{for} \quad T<T_c.$$

By the translational invariance of $\bar{\rho}$, this is equivalent to

$$\sum_{x\in X} \bar{\rho}(\hat{\eta}(x)\,.\,\hat{\eta}(0))=\infty \quad \text{for} \quad T<T_c.$$

We assume that this condition persists, by continuity, to the critical point, so that the unique equilibrium state there, ρ_c, satisfies the condition

$$\sum_{x\in X} \rho_c(\hat{\eta}(x)\,.\,\hat{\eta}(0))=\infty. \tag{6}$$

Since ρ_c is a pure phase for which $\rho_c(\hat{\eta}(x))=0$, by symmetry, this condition specifies a precise sense in which the fluctuations of the order field have infinite-range correlations. In fact, the condition (6) is the basic assumption of the theory of critical phenomena,† and it represents an incipient macroscopic degeneracy, or long-range order, at the critical point. In simple cases, e.g. when Σ is an Ising ferromagnet and $\hat{\eta}(x)$ the spin at the site x, it corresponds to the divergence of the susceptibility at the critical point.‡ In general, the assumption of condition (6) is amply supported by the empirical fact of critical opalescence.§ We remark that this condition is satisfied if $\rho_c(\hat{\eta}(x)\,.\,\hat{\eta}(0))\sim|x|^{-l}$ at large distances, with l less than the dimensionality d, but not if this correlation function falls off exponentially with distance.

Off-diagonal long-range order. The ordering we have discussed so far is connected with symmetry breakdown. A different kind of ordering was conceived by Yang (1962) in connection with the theory of superfluidity. Drawing on ideas proposed earlier by Penrose and Onsager (1956), Yang observed that the microscopic theories of superfluid states were encapsulated by the assumption of long-range correlations between *quantized fields* representing the particles of the superfluids. In the case of a Bose superfluid, such as He^4, this assumption takes the form

$$\langle\psi^*(x)\psi(x')\rangle\to\nu \quad (>0) \quad \text{as} \quad |x-x'|\to\infty \tag{7}$$

† See Fisher (1967*b*).

‡ For an Ising ferromagnet, one can show this by formulating the susceptibility of a finite version of the model and then using Griffiths's (1967) correlation inequalities to show that the susceptibility of the infinite system at the critical point is $(kT_c)^{-1}$ times the LHS of (6). More generally, when one lacks such inequalities, it is not always clear that condition (6) represents the divergence of a critical susceptibility.

§ See Fisher (1967*b*).

for equilibrium states below the transition temperature, ψ being the quantum field representing the particles. In the case of a Fermi superfluid, such as the system of electrons in a superconductor, the corresponding condition is obtained by replacing the ψ in this formula by $\psi_\uparrow \psi_\downarrow$, the product of the quantum fields for the particles with 'up' and 'down' spins. The property represented by (7), or its counterpart for a Fermi system, is generally referred to as *off-diagonal long-range order* (*ODLRO*), following Yang. This is different from the standard kind of ordering, considered above, because ψ (or $\psi_\uparrow \psi_\downarrow$) is not an observable, as it changes under gauge transformations of the first kind, $\psi \to \psi\, e^{i\alpha}$. Moreover, as all states of the system are gauge invariant, irrespective of whether or not they are pure phases, the gauge dependence of the quantum field ensures that its expectation value is always zero and therefore cannot serve as an order parameter.

In fact, the standard and natural choice for the order parameter in the case of ODLRO is the variable v, representing the strength of the long-range correlations of the quantum field in (7). For a Bose system, this parameter corresponds to the density of particles with zero momentum, i.e. the condensate density, since it follows from (7) that $v = \langle a^*(0)a(0)\rangle$, where $a(k)$ is the Fourier transform of $\psi(x)$. In the case of a superconductor, v corresponds to the density of Cooper pairs.

Note on macroscopic wave functions. Under conditions of spatial inhomogeneity, which prevail in various non-equilibrium situations, the ODLRO assumption (7) may be generalized to the form

$$\langle \psi^*(x)\psi(x')\rangle - \Psi^*(x)\Psi(x') \to 0 \quad \text{as } |x - x'| \to \infty,$$

the $\Psi(x)$ being determined up to a constant phase factor. Here, $\Psi(x)$ is the so-called macroscopic wave function of the Ginzburg–Landau theory† and the space-dependent phase of this function serves as a velocity potential in the Josephson tunnelling effect‡ and, more generally, in superfluid hydrodynamics.§

Further comments on ordering and long-range correlations. The picture we have arrived at is that ordering corresponds to long-range correlations, whether of a gauge field, $\psi(x)$, or of an observable field, $\hat{\eta}(x)$, that transforms under the action of a local symmetry group. This connection between ordering and long-range correlations can certainly be extended to systems that exhibit neither symmetry breakdown nor ODLRO. For example, a real liquid–gas transition has a macroscopic degeneracy below the critical point, with the pure liquid and gaseous phases having different

† See Ginzburg and Landau (1950), and Ginzburg (1955).
‡ See Josephson (1964).
§ See Weiss and Haug (1973).

densities and microscopic structures, without apparently any symmetry breakdown or ODLRO. For this system, the difference between the densities of the two phases at the same temperature serves as an order parameter, since it varies continuously below the transition point and vanishes there. The ordering, represented by this parameter, in the liquid and gaseous phases corresponds to the excess and deficit, respectively, of particles per unit volume, relative to the density of the mixed phase given by their arithmetic mean. Furthermore, it follows from the argument leading to (2) that the density fluctuations in this mixed phase will have long-(infinite)-range correlations by virtue of the difference between the densities of the pure phases.

This argument is quite general for systems that undergo second-order phase transitions with macroscopic degeneracy below the transition point, and so we conclude that these transitions always involve an ordering that corresponds to some long-range correlations.

5.3. Criteria for the existence of order–disorder transitions

The question of whether a given system undergoes an order–disorder transition reduces to two basic problems. The first is whether it exhibits ordering, as represented by non-zero values of some order parameter, at sufficiently low temperatures: the second is whether it has no such ordering at sufficiently high ones. In fact, the latter problem has been resolved for systems with short-range forces† and also for some Coulomb systems,‡ by a rigorous version of Mayer's (1947) classic cluster expansions that reveal that these systems have no ordering, i.e. long-range correlations, at sufficiently high temperatures. Since these results cover the models we shall be concerned with here, we shall take it that the problem of whether a system undergoes an order–disorder transition reduces to that of whether it exhibits low-temperature ordering. Our objective here is to discuss the picture that has emerged from the solution of this problem for various classes of systems. We shall start, in §5.3.1, by presenting some simple energy-versus-entropy arguments that demonstrate the key roles played by dimensionality, symmetry, and interaction range in order–disorder transitions and establish that one-dimensional systems, as well as certain two-dimensional ones, with short-range forces do not support ordered phases. We shall then turn to the methodology mentioned briefly in §5.1, that has been devised to prove the existence of order–disorder transitions of both the discrete and continuous symmetry-breaking kinds. Thus, after specifying, in §5.3.2, some of the positive results that this has provided, we shall give a fairly detailed account, in §5.3.3, of the argument by J. Fröhlich, Simon, and Spencer

† See Ginibre (1965, 1968) and Gallavotti, Miracle-Sole, and Robinson (1968).
‡ See Brydges and Federbush (1980).

(FSS) (1976), proving the existence of an order–disorder transition, with continuous rotational symmetry breakdown, in the classical Heisenberg ferromagnet. We shall then describe in outline how the extended forms of the Peierls and FSS arguments form the basis of a methodology for proving the existence of low temperature ordering in a rather wide class of models.

It will be seen that this methodology is largely dependent on the systems concerned possessing a certain symmetry property with respect to space reflections. This is the property of *reflection positivity* (RP), which arises in quantum field theory† and may be described as follows. Suppose that Σ is a system in a space X, consisting of two halves, X_1 and X_2, that are mirror images of one another in a plane, Π. Then one can generally define an operator, θ, representing reflections on Π, that maps the observables of X_1 into those of X_2. A state is termed RP if its expectation values satisfy the condition

$$\langle A\theta A\rangle \geqslant 0 \tag{8}$$

for all observables A of X_1. It turns out that this is a common, though by no means universal, property of canonical equilibrium states.‡ Its value, when applicable, is that it permits drastic simplifications of the estimates needed to carry through the FSS and generalized Peierls arguments.

5.3.1. The absence of order–disorder transitions in some low-dimensional models: energy-versus-entropy arguments

Since equilibrium states correspond to the minima of the free energy functional $\hat{e} - T\hat{s}$, it is evident that their properties are governed by the competition between the demands of low energy and high entropy, the balance between these demands being determined by the temperature. Landau§ provided a simple argument, based on energy-versus-entropy estimates, to the effect that one-dimensional systems with short-range interactions cannot undergo phase transitions. This argument has subsequently been made rigorous and also extended to prove a similar result for some two-dimensional models with *continuous* symmetries, by Simon and Sokal (1981): by contrast, of course, the two-dimensional Ising model, whose internal symmetry (spin reflections) is *discrete*, does undergo an order–disorder transition. So, too, does a one-dimensional system of Ising spins, with suitable long-range interactions, as has been proved by Dyson (1969). Thus, it is clear that dimensionality, symmetry, and interaction range are key factors in determining which systems undergo order–disorder transitions.

† In quantum field theory, RP is a key property of the analytic continuation to imaginary times of the space–time correlations of fields in the vacuum state (cf. Osterwalder and Schrader 1973, 1975).

‡ See J. Fröhlich, Israel, Lieb, and Simon (1978, 1980).

§ See Landau and Lifshitz (1959), p. 482.

We shall now sketch the argument of Simon and Sokal (1981) for the non-existence of order–disorder transitions in certain low-dimensional models, so as to see how these factors come into play. In the case of a one-dimensional Ising system with nearest-neighbour, or finite-range, ferromagnetic interactions, the argument proceeds as follows. Suppose that ρ is a positively polarized state, which we take to be a candidate for thermal equilibrium at temperature T. Then we may examine the stability of ρ against modifications that correspond to the introduction of negatively polarized regions by estimating the resultant changes in energy and entropy. Thus, we start by reversing the spins in a finite segment, Δ, of length N, which we take to be 'large' – this amounts to the introduction of a macroscopic, negatively polarized 'droplet'. The resultant state ρ' is given by $\rho'(A) \equiv \rho(\tau_\Delta A)$, where τ_Δ is the spin reversal automorphism for the segment Δ. In a similar way, we may define states $\rho_1', \ldots, \rho_n'$ obtained from ρ by reversing the spins in adjacent segments $\Delta_1, \ldots, \Delta_n$, respectively, each of length N. Thus, the state

$$\tilde{\rho} = n^{-1}(\rho_1' + \ldots + \rho_n'),$$

corresponds to the modification of ρ due to the insertion of a negatively polarized droplet of length N, which may take up any of the positions $\Delta_1, \ldots, \Delta_n$, each with probability n^{-1}. The introduction of the droplet, with this choice of positions, introduces an element of randomness, leading to an entropy increase†

$$\Delta S \simeq k \ln n \tag{9}$$

for fixed n and large N. On the other hand, the change in energy due to the introduction of the droplet is bounded by a constant, independent of N, as the interactions are of finite range. Consequently, the increment free energy is (const. $- kT \ln n$), which is negative for sufficiently large n. This implies that ρ is not locally stable‡ and therefore is not an equilibrium state. Hence the model cannot support any ordering corresponding to polarization at finite temperatures.

Comment. This argument evidently depends crucially on both the one-dimensionality of the model and the finite range of the interactions. For if, in a one-dimensional Ising system, the pair interaction potential were of long range, falling off as (distance)$^{-\alpha}$, with $1 < \alpha < 2$, as in Dyson's (1969) models, then the energy increment would be

$$\Delta E \sim \text{const. } N^{2-\alpha} \tag{10a}$$

and this would offset the entropy gain (9) for large N. A similar thing would

† In fact, although the estimate (9) for ΔS was obtained by Simon and Sokal (1981) for arbitrarily large, finite systems, there is no difficulty in extending their result to infinite systems, where the incremental entropy ΔS is defined as in §3.4.3.

‡ Cf. §§3.4 and 3.6.

happen for a d-dimensional Ising system ($d \geq 2$) with short range interactions. For if one randomly inserted a cubic 'wrongly polarized droplet' of volume N in one or another of n contiguous regions, whose union formed a cube of volume nN, then for fixed n and large N, the entropy gain would still be $\simeq k \ln n$, while the energy cost would be proportional to the surface area of the cube, i.e.

$$\Delta E \sim \text{const.}\, N^{(d-1)/d} \tag{10b}$$

and so would offset the entropy increase. Evidently, the formulae (10a,b) indicate that long-range interactions (small α) and high dimensionality both favour the stabilization of a polarized state against the insertion of oppositely polarized droplets.

We now turn to another model, namely the two-dimensional system of classical plane rotators with nearest-neighbour interactions of the form $-J\cos(\theta - \theta')$, where $J > 0$ and θ and θ' are the inclinations of the rotators to some fixed direction.† This is a model whose internal symmetry is *continuous*, corresponding as it does to rotations $\theta \to \theta + \text{const}$. What we shall now see is that this continuous symmetry is less favourable for ordering than the discrete one of the Ising model and that, in fact, the two-dimensional plane rotator system cannot support polarized equilibrium states. The argument runs as follows (cf. Simon and Sokal 1981). Suppose that ρ, a candidate for an equilibrium state of the model, is polarized along the direction Ox, say. Then a relatively economical way of inserting an oppositely polarized droplet may be achieved as follows. Starting from the polarized state ρ, reverse the spins in a square, Δ, of area N, i.e. of side $L = N^{\frac{1}{2}}$. Then rotate the spins in the L successive square-shaped 'shells' of unit thickness surrounding Δ in such a way that the spins in the innermost shell are rotated through $\pi(1 - L^{-1})$, those in the next shell by $\pi(1 - 2L^{-1})$, etc. The angles of relative rotation between successive shells, as well as those between the innermost shell on Δ and the outermost one and the region beyond it, is then πL^{-1}. The change in interaction energy between neighbouring shells is then proportional to $L(1 - \cos \pi L^{-1})$, and consequently the total energy required to introduce the droplet comprising the negatively polarized region Δ and the L shells surrounding it is given essentially by

$$\Delta E \sim \text{const.}\, L^2 . (1 - \cos \pi L^{-1}) \sim \text{const.}, \quad \text{for large } L.$$

Note that this contrasts with corresponding estimate, given by const. $N^{\frac{1}{2}}$, for the two-dimensional Ising system. Further, the estimate (9) for the entropy given due to the random insertion of the droplet in one or other of n equivalent places is still applicable: here, the regions where the droplet is to

† For this model, a spin configuration corresponds to the values of the angles, θ_x, representing the inclination to some fixed direction of the rotators at the various sites, x; and the counterpart of $\text{Tr}_\Lambda(\ldots)$ is $\prod_{x \in \Lambda} \int_0^{2\pi} d\theta_x(\ldots)$.

be placed can be taken to be n squares, each of side $3L$, whose centres form a square array. It follows now from these estimates for the energy and entropy changes that the increment in free energy due to the random insertion of the droplet $\sim$ (const. $- kT \ln n$), as in the case of the one-dimensional Ising model. Since this incremental free energy is negative for sufficiently large n, we conclude that the two-dimensional plane rotator cannot support polarized equilibrium states at finite temperatures. It cannot therefore undergo an order–disorder transition, where the ordered phase is polarized.

The results we have discussed here on the basis of simple energy-versus-entropy estimates have been extended by rather more refined methods to other low-dimensional systems. In particular, Araki (1969) has proved the absence of phase transitions in one-dimensional systems with finite-range interactions by an analysis based on the KMS condition, while Mermin and Wagner (1966) have proved the absence of polarized equilibrium states in the two-dimensional Heisenberg ferromagnet by an argument based on correlation inequalities† that ensue from that condition. We remark that, in view of the equivalence between local stability and the KMS condition, noted in §3.7, these arguments have the same physical content as those based on energy and entropy estimates.

5.3.2. Models with order–disorder transitions

The Peierls argument,‡ that the two-dimensional Ising ferromagnet has polarized equilibrium states at sufficiently low temperatures, implies that this model undergoes an order–disorder transition with spin reversal symmetry breakdown. This argument has been extended to prove the same result, in two or more dimensions, for systems of Ising spins§ with ferromagnetic interactions, that may couple non-nearest neighbours, and also for the *highly anisotropic* Heisenberg model¶ with short-range interactions between spins σ and σ' of the form

$$-J(\alpha\sigma_x\sigma'_x + \alpha\sigma_y\sigma'_y + \sigma_z\sigma'_z), \tag{11}$$

with J and α positive and α sufficiently small ($\ll 1$). In this latter model, the low-temperature spontaneous polarization is along $\pm Oz$.

These results have been obtained from essentially the original form of the Peierls argument, without recourse to the RP condition (8). The following systems, however, are among those that have been proved to undergo order–disorder transitions by the methods, which we shall presently formulate, that utilize the RP condition in an essential way.

† The particular inequalities employed there are due to Bogoliubov (1962). As proved by Fannes and Verbeure (1977), they are just one of a class of inequalities that follow from the KMS condition.

‡ Cf. §3.5.1.

§ See Griffiths (1967).

¶ See Ginibre (1969) and Robinson (1969).

Systems with breakdown of discrete symmetry

(a)† The anisotropic Heisenberg antiferromagnet, with spins $\geqslant 1$, in two or more dimensions. This is the model with nearest-neighbour couplings of the form (11), though now with $J < 0$ and α taking arbitrary positive values less than unity. The model exhibits low-temperature ordering, in which the subsystems of spins on two complementary, interlocking sublattices are equally and oppositely polarized along the $\pm Oz$ directions. The ordering therefore breaks the symmetry corresponding to reversals of the z-components of the spins.

(b)‡ An electrically neutral system, consisting of two species of particles with equal and opposite charges and interacting via Coulomb forces, on a three-dimensional lattice. This system exhibits low-temperature ordering with the two species of particles preferentially occupying complementary sublattices, so as to provide an ionic crystalline configuration. There is thus a breakdown of the charge conjugation symmetry, corresponding to the interchange of equal and opposite charges.

Systems with breakdown of continuous symmetry

(c)§ The classical three-dimensional Heisenberg ferromagnet: this is a system of classical vector spins, σ, of unit modulus, on a simple cubic lattice, with nearest neighbour couplings of the form $-J\sigma \cdot \sigma'$, J being positive. This model exhibits low-temperature ordering given by spontaneous polarization, which breaks the rotational symmetry of the interactions.

(d)¶ The three-dimensional (quantum) XY-model, consisting of a system of Pauli spins on a three-dimensional lattice, with nearest-neighbour couplings of the form $-J(\sigma_x\sigma_x' + \sigma_y\sigma_y')$ and $J > 0$. This model exhibits an order–disorder transition, with low-temperature ordering corresponding to a polarization along some direction in the xy-plane, which breaks the symmetry of rotations of the spins in that plane.

(e)|| The three-dimensional (quantum) Heisenberg antiferromagnet, consisting of spins on a three-dimensional lattice with nearest neighbour couplings of the form $J\sigma \cdot \sigma'$ and $J > 0$. In the case where the spins $\geqslant 1$, the model exhibits low-temperature ordering, corresponding to equal and opposite polarizations of two complementary sublattices. This breaks the symmetry of the rotations of the spins.

† See J. Fröhlich and E. H. Lieb (1978).
‡ See J. Fröhlich, Israel, Lieb, and Simon (1980).
§ See J. Fröhlich, Simon, and Spencer (1976).
¶ See Dyson, Lieb, and Simon (1978).
|| See Dyson, Lieb, and Simon (1978).

Note. It has *not* as yet been proved that the three-dimensional quantum Heisenberg ferromagnet undergoes an order–disorder transition, the difficulty being that the RP arguments on which the present methods are based are inapplicable to that model. This point will be discussed further in §5.3.4.

5.3.3. The FSS proof of ordering in the classical Heisenberg ferromagnet

The Model. The classical Heisenberg ferromagnet is a system, Σ, of spin vectors, $\sigma(x)$, placed at the sites, x, of a three-dimensional, simple cubic lattice of unit spacing. It is assumed that the spin vectors are of unit modulus, i.e. $|\sigma(x)| = 1$, and that the interactions are confined to nearest neighbours and take the form $-J\sigma(x) \,.\, \sigma(x')$, with J a positive constant.

We shall relate the properties of Σ to those of a finite version $\Sigma(\Lambda)$ of this model, comprising the spins in a cubic region Λ, with *periodic* boundary conditions. Thus, denoting by $\delta_1, \delta_2, \delta_3$ the unit vectors corresponding to displacements between neighbouring sites along the crystallographic axes, it follows from the above specifications that the Hamiltonian for $\Sigma(\Lambda)$ takes the form

$$-J\sum_{i=1}^{3}\sum_{x\in\Lambda} \sigma(x) \,.\, \sigma(x+\delta_i),$$

$\sigma(x)$ having the periodicity of the cube. Since $|\sigma(x)| = 1$, we may re-express this Hamiltonian in the following form after adding a constant:

$$H(\Lambda) = \tfrac{1}{2}J \sum_{i=1}^{3}\sum_{x\in\Lambda} (\sigma(x+\delta_i) - \sigma(x))^2. \tag{12}$$

The phase space for each spin $\sigma(x)$ is the unit sphere. Assuming the directional distribution for a *free* spin, σ, to be isotropic, the probability of finding the direction of σ in a cone of solid angle $\mathrm{d}\Omega(\sigma)$ with vertex at the origin is $(4\pi)^{-1}\,\mathrm{d}\Omega(\sigma) \equiv \mathrm{d}\omega(\sigma)$. The canonical equilibrium expectation value of an observable A of Σ is therefore

$$\langle A\rangle_\Lambda = \langle A\,\mathrm{e}^{-\beta H(\Lambda)}\rangle_{0,\Lambda}/\langle \mathrm{e}^{-\beta H(\Lambda)}\rangle_{0,\Lambda} \tag{13}$$

where

$$\langle \ldots\rangle_{0,\Lambda} = \int \ldots \prod_{x\in\Lambda} \mathrm{d}\omega(\sigma(x)). \tag{14}$$

Order as condensation of spin waves. The FSS argument is centred on an estimate for the dispersion $m(\Lambda)$ in the magnetic moment of $\Sigma(\Lambda)$, as defined by the formula

$$m(\Lambda) = |\Lambda|^{-1}\left\langle\left(\sum_{x\in\Lambda} \sigma(x)\right)^2\right\rangle_\Lambda^{1/2}. \tag{15}$$

For it follows from Griffiths's theorem, which we discussed in §5.2 (cf. eqn (3)), that if $m(\Lambda)$ tends to a non-zero limit as Λ increases to infinity, then the polarization of the pure equilibrium phases of the *infinite system* Σ is guaranteed.

The treatment of $m(\Lambda)$ is based on a spin wave description of $\Sigma(\Lambda)$. In a usual way, a spin wave of wave-vector q is defined by the standard formula

$$\hat{\sigma}(q) = |\Lambda|^{-\frac{1}{2}} \sum_{x \in \Lambda} \sigma(x)\, \mathrm{e}^{-\mathrm{i}q \cdot x}, \tag{16}$$

q taking values for which $e^{\mathrm{i}q \cdot x}$ has the periodicity of the box Λ. Since $\sigma^2(x) \equiv 1$, it follows from (16) that

$$\sum_q |\hat{\sigma}(q)|^2 = |\Lambda|. \tag{17}$$

Likewise, it follows from (12) and (16) that

$$H(\Lambda) = \text{const.} + \sum_q E(q)\, |\hat{\sigma}(q)|^2 \tag{18}$$

where

$$E(q) = J \sum_{i=1}^{3} (1 - \cos q \,.\, \delta_i), \tag{19}$$

the spin wave energy for wave-vector q.

By eqns (16) and (17), the formula (15) for $m(\Lambda)$ may be expressed as

$$m(\Lambda)^2 = |\Lambda|^{-1} \langle \hat{\sigma}(0)^2 \rangle_\Lambda = 1 - |\Lambda|^{-1} \sum_{q \neq 0} \langle |\hat{\sigma}(q)|^2 \rangle_\Lambda. \tag{20}$$

Hence, in order to establish the desired result, that the infinite volume limit of $m(\Lambda)$ is non-zero, it suffices to prove that the magnitude of the last term in this formula has a Λ-independent upper bound B, that is less than unity, i.e. that

$$|\Lambda|^{-1} \sum_{q \neq 0} \langle |\hat{\sigma}(q)|^2 \rangle_\Lambda < B < 1 \tag{21}$$

for all sufficiently large Λ. We remark that, in view of eqns (17) and (18), this inequality is analogous to the condition for Bose–Einstein condensation in an ideal Bose gas, with $|\hat{\sigma}(q)|^2$ and $E(q)$ playing the roles of the number of particles and energy per particle with wave-vector q, $|\Lambda|$ corresponding to the total number of particles and $|\Lambda|^{-1} \langle \hat{\sigma}(0)^2 \rangle_\Lambda$ to the density of the condensate.

Infra-red bound. The FSS approach to the proof of the condensation inequality (21) is to show that the spin wave energy $E(q) \langle |\hat{\sigma}(q)|^2 \rangle_\Lambda$, for $q \neq 0$, does not exceed the classical equipartition value, $\frac{3}{2}kT$, i.e. that

$$\langle |\hat{\sigma}(q)|^2 \rangle_\Lambda \leqslant \frac{3kT}{2E(q)} \quad \text{for} \quad q \neq 0. \tag{22}$$

To see that this inequality leads to (21), we note that it implies that the LHS of that formula is majorized by

$$\tfrac{3}{2}kT\,|\Lambda|^{-1}\sum_{q\neq 0}(E(q))^{-1},$$

which may be approximated, to adequate accuracy (for large Λ), by

$$\tfrac{3}{2}kT\frac{1}{(2\pi)^3}\int\frac{d^3q}{E(q)},$$

the integration being taken over the cube $|q_i|\leqslant\pi$ for $i=1,2,3$. Since, by (19), $E(q)$ is strictly positive for $q\neq 0$ and $\simeq\tfrac{1}{2}Jq^2$ for small $|q|$, the integral in this last expression is a finite constant, which means that the LHS of (21) is majorized by const. T, and is therefore less than unity at sufficiently low temperatures. This confirms that the inequality (22) guarantees the fulfillment of the condensation condition (21) when the temperature is low enough. The bound on $\langle|\hat{\sigma}(q)|^2\rangle_\Lambda$ given by (22) is usually referred to as an *infra-red bound*, since its essential significance lies in the limitation it imposes on the divergence of $\langle|\hat{\sigma}(q)|^2\rangle_\Lambda$ for small $|q|$.

Thus, the proof of low-temperature ordering of the model may be reduced to that of the infra-red bound (22).

Response to perturbations. The proof of the infra-red bound may in turn be formulated in terms of the response of $\Sigma(\Lambda)$ to suitable external perturbations. For suppose that this system is perturbed in such a way that the energy of interaction between spins at neighbouring sites is changed from $\tfrac{1}{2}J(\sigma(x)-\sigma(x'))^2$, as in (12), to $\tfrac{1}{2}J(\sigma(x)-\sigma(x')-h(x,x'))$, where $h\equiv\{h(x,x')=-h(x',x)\}$ corresponds to an external field attached to the bonds linking nearest neighbours. The Hamiltonian of the perturbed system is then (cf. (12))

$$\begin{aligned}H(\Lambda,h)&=\tfrac{1}{2}J\sum_{i=1}^{3}\sum_{x\in\Lambda}(\sigma(x)-\sigma(x+\delta_i)-h(x,x+\delta_i))^2\\&\equiv\tfrac{1}{2}J\sum{}'(\sigma(x)-\sigma(x')-h(x,x'))^2,\end{aligned}\tag{23}$$

the prime over the last Σ indicating that summation is confined to nearest neighbours, with each pair counted once. The free energy of the perturbed system is

$$F_\Lambda(h)=-kT\ln Z_\Lambda(h),\tag{24}$$

where

$$Z_\Lambda(h)=\langle\exp(-\beta H(\Lambda,h))\rangle_{0,\Lambda},\tag{25}$$

with $\langle\ldots\rangle_{0,\Lambda}$ defined by (14).

We shall now show that the problem of proving the infra-red bound can be reduced to that of showing that $F_\Lambda(h)$ is absolutely minimized at $h=0$. The argument runs as follows. By (13), (24), and (25),

$$F_\Lambda(h)=F_\Lambda(0)-kT\ln\langle\exp(\beta J\,\partial\sigma(h))\rangle_\Lambda+\tfrac{1}{2}J\,|h|^2, \tag{26}$$

where

$$\partial\sigma(h)=\sum_{i=1}^{3}\sum_{x\in\Lambda}(\sigma(x)-\sigma(x+\delta_i))\,.\,h(x,x+\delta_i), \tag{27}$$

and

$$|h|^2=\sum_{i=1}^{3}\sum_{x\in\Lambda}|h(x,x+\delta_i)|^2\equiv\Sigma'\,|h(x,x')|^2. \tag{28}$$

Now the minimization of $F_\Lambda(h)$ at $h=0$ would imply that, for fixed h, the function $F_\Lambda(\lambda h)$ of the real variable λ would be minimized at $\lambda=0$, and therefore that

$$\frac{\mathrm{d}}{\mathrm{d}\lambda}F_\Lambda(\lambda h)=0 \quad\text{and}\quad \frac{\mathrm{d}^2}{\mathrm{d}\lambda^2}F_\Lambda(\lambda h)\geqslant 0 \quad\text{at}\quad h\geqslant 0,$$

or, equivalently, by (26)–(28), that

$$\langle\partial\sigma(h)\rangle_\Lambda=0 \tag{29}$$

and

$$\beta J\langle|\partial\sigma(h)|^2\rangle_\Lambda\leqslant|h|^2, \tag{30}$$

the expectation values here being for the unperturbed system. In fact, (29) follows trivially from the spin reversal symmetry of the canonical state. The condition (30), however, as applied for arbitrary h, is non-trivial. Further, it may be extended to complex $h=h^{(1)}+\mathrm{i}h^{(2)}$ (with $h^{(1)}$, $h^{(2)}$ real) by adding together the versions of that inequality for $h^{(1)}$ and $h^{(2)}$. On choosing

$$h(x,x')=|\Lambda|^{-\frac{1}{2}}(\exp\mathrm{i}q\,.\,x-\exp\mathrm{i}q\,.\,x')u, \quad\text{with}\quad q\neq 0,$$

with u an arbitrary unit vector, it follows from (16), (19), (27), and (28) that the inequality (30) reduces to the form

$$\langle|\hat{\sigma}(q)\,.\,u|^2\rangle_\Lambda\leqslant kT/2E(q), \quad\text{for}\quad q\neq 0;$$

and the required infra-red bound (22) follows from adding together the versions of this last formula corresponding to three mutually orthogonal values of u. We conclude therefore that the absolute minimization of F_Λ at $h=0$ provides a sufficient condition for the validity of this bound. Hence, *the whole problem of proving the low-temperature ordering of the model reduces to that of showing that $F_\Lambda(h)$ is absolutely minimized at $h=0$.*

We note here that it follows from the definition of $F_\Lambda(h)$ that this function does attain an absolute minimum, and does not merely approach a lower bound arbitrarily closely, for the following reasons. By (26)–(28), $F_\Lambda(h)$ is a continuous function of the components, $h(x, x')$, of h and tends to infinity like $\frac{1}{2}J\,|h|^2$ with $|h|$. This guarantees that, for some positive constant C, $F_\Lambda(h) > F_\Lambda(0)$ when $|h| > C$. Hence, as the continuity of F_Λ ensures† that its restriction to the closed bounded region $|h| \leqslant C$ attains an absolute minimum, it follows that the unrestricted function F_Λ does so, too. The problem that remains is to prove that the minimum of F_Λ is reached at $h = 0$.

Reflection positivity. It is at this stage that the reflection positivity property becomes pertinent. In order to specify this property for the finite system $\Sigma(\Lambda)$, we consider a division of Λ into two parts, Λ_1 and Λ_2, that are mirror images of one another in a plane, Π, that contains no lattice sites and is perpendicular to one of the principal crystallographic directions. Thus, for example, if Λ consists of the sites $x = (x_1, x_2, x_3)$, whose components run from 1 to $2N$, then Π could be taken to be the plane $x_i = N + \frac{1}{2}$ (for $i = 1, 2, 3$), in which case Λ_1 and Λ_2 would be the two halves of Λ for which $x_i \leqslant N$ and $x_i \geqslant N + 1$, respectively. Moreover, as Λ is a periodic box, one could equally well choose Π to be the plane $x_j = n + \frac{1}{2}$, for any integer n, in which case Λ_1 and Λ_2 would be the subregions of Λ for which $x_j = n, n-1, \ldots, n-N \pmod{2N}$‡ and $x_j = n, n+1, \ldots, n-N \pmod{2N}$, respectively.

In general, if x is a site of Λ_1, we shall denote its mirror image in Π by $\tilde{x}$. We define the algebras of observables, $\mathscr{A}_1$ and $\mathscr{A}_2$, for Λ_1 and Λ_2, to be the respective functions of the spins in those regions, and the reflection operator θ to be the transformation of $\mathscr{A}_1$ into $\mathscr{A}_2$ given by

$$\theta F(\sigma(x), \sigma(x'), \ldots, \sigma(x^{(l)})) \equiv \overline{F(\sigma(\tilde{x}), \sigma(\tilde{x}'), \ldots, \sigma(\tilde{x}^{(l)}))}. \tag{31}$$

This operator is therefore antilinear, i.e.

$$\theta(\lambda A + \mu B) = \bar{\lambda}\theta A + \bar{\mu}\theta B \tag{32}$$

for complex numbers λ, μ and elements A, B of $\mathscr{A}_1$. It follows now from (14) and (31) that

$$\langle A\theta A\rangle_{0,\Lambda} = |\langle A\rangle_{0,\Lambda}|^2 \text{ for } A \text{ in } \mathscr{A}_1,$$

and therefore condition $\langle \ldots \rangle_{0,\Lambda}$ satisfies the *reflection positivity* (RP) condition

$$\langle A\theta A\rangle_{0,\Lambda} \geqslant 0 \text{ for } A \text{ in } \mathscr{A}_1. \tag{33}$$

Hence, by the antilinearity of θ, $\langle A\theta B\rangle_{0,\Lambda}$ corresponds to a Hilbert space inner product (A, B), and therefore the following Schwartz inequality holds.

$$|\langle A\theta B\rangle_{0,\Lambda}| \leqslant (\langle A\theta A\rangle_{0,\Lambda}\langle B\theta B\rangle_{0,\Lambda})^{\frac{1}{2}}. \tag{34}$$

† Cf. the argument of §3.2.

‡ In a usual way, $m \pmod{2N} = m - 2Nk$ if m lies between $2Nk$ and $2N(k+1)$, with k an integer.

Suppose now that, for any field h, the fields h_1 and h_2 are defined so that (a) they coincide with h in Λ_1, Λ_2, respectively, (b) they vanish on the bonds connecting Λ_1 to Λ_2, and (c) they are invariant under the reflections $x, x' \to \bar{x}, \bar{x}'$, i.e.

$$h_j(x, x') = h(x, x') \quad \text{for } x, x' \text{ both in } \Lambda_j, j = 1, 2,$$

$$h_j(x, x') = 0 \quad \text{for } x \text{ in } \Lambda_1 \text{ and } x' \text{ in } \Lambda_2, \text{ or vice versa,} \tag{35}$$

$$h_j(x, x') = h_j(\bar{x}, \bar{x}') \quad \text{for } x, x' \text{ in } \Lambda_1.$$

Then, as we shall presently show, one can use the RP inequality (34) and the definition (25) of $Z_\Lambda(h)$ to prove that

$$Z_\Lambda(h) \leqslant (Z_\Lambda(h_1)Z_\Lambda(h_2))^{\frac{1}{2}}, \tag{36}$$

or equivalently, by (24),

$$F_\Lambda(h) \geqslant \tfrac{1}{2}(F_\Lambda(h_1) + F_\Lambda(h_2)). \tag{37}$$

This is the key inequality for the proof that F_Λ is minimized at $h = 0$. The proof runs as follows. Suppose that F_Λ takes its minimum value, $F_\Lambda^{(\min)}$, when $h = \bar{h}$, say. Then, by (37),

$$F_\Lambda^{(\min)} = F_\Lambda(\bar{h}) \geqslant \tfrac{1}{2}(F_\Lambda(\bar{h}_1) + F_\Lambda(\bar{h}_2)),$$

and hence, as neither $F_\Lambda(\bar{h}_1)$ nor $F_\Lambda(\bar{h}_2)$ can be less than $F_\Lambda^{(\min)}$, it follows that

$$F_\Lambda(\bar{h}_1) = F_\Lambda(\bar{h}_2) = F_\Lambda^{(\min)}. \tag{38}$$

Now if $\bar{h} \neq 0$, it must have a positive number, l, of non-zero components. In this case, one can choose the plane Π so that it intersects at least one bond on which $\bar{h}$ is non-zero, thereby ensuring, by (35), that if l_1 and l_2 are the numbers of bonds on which $\bar{h}_1$ and $\bar{h}_2$, respectively, do not vanish, then

$$l_1 + l_2 < 2l,$$

and hence at least one of the numbers (l_1, l_2) is less than l. Therefore, by (38), if F_Λ is minimized by a value $\bar{h}$ of h with l (>0) non-zero components, it is also minimized by another value of h, namely $\bar{h}_1$ or $\bar{h}_2$, with fewer than l non-zero components. From this, it follows by induction that $F_\Lambda(h)$ is minimized at $h = 0$, as required.

It therefore now remains for us to prove the inequality (36).

Derivation of the inequality (36) *from RP*. By eqns (23) and (35), the Hamiltonian $H(\Lambda, h)$ may be expressed in the form

$$H(\Lambda, h) = H(\Lambda_1, h_1) + H(\Lambda_2, h_2) + H_I(h), \tag{39}$$

where

$$H(\Lambda_j, h_j) = \tfrac{1}{2}J \sum_{x,x' \in \Lambda_j} (\sigma(x) - \sigma(x') - h_j(x, x'))^2 \tag{40}$$

is the partial Hamiltonian for Λ_j ($j = 1, 2$) and

$$H_I(h) = \tfrac{1}{2}J \sum_{\substack{x \in \Lambda_1 \\ x' \in \Lambda_2}} (\sigma(x) - \sigma(x') - h(x, x'))^2$$

is the coupling between Λ_1 and Λ_2. Since this coupling involves only the bonds connecting the plane Π_1 of Λ_1, nearest and parallel to Π_1 with its reflection Π_2 in Π, we may rewrite this last formula as

$$H_I(h) = \tfrac{1}{2}J \sum_{x \in \Pi_1} (\sigma(x) - \sigma(\tilde{x}) - h(x, \tilde{x}))^2, \tag{41}$$

or, in shorthand notation,

$$H_I(h) = \tfrac{1}{2}J(\sigma - \tilde{\sigma} - h_I)^2, \quad \text{with} \quad \tilde{\sigma} = \theta\sigma. \tag{41$'$}$$

By (25) and (39), $Z_\Lambda(h)$ may be expressed in the form

$$Z_\Lambda(h) = \langle \exp(-\beta H(\Lambda_1, h_1)) \exp(-\beta H_I(h)) \exp(-\beta H(\Lambda_2, h_2)) \rangle_{0,\Lambda}. \tag{42}$$

Now, in view of the Fourier representation of Gaussian functions, i.e.

$$\exp(-\tfrac{1}{2}\alpha x^2) = \text{const.} \int \exp\left(-\frac{k^2}{2\alpha} - \mathrm{i}kx\right) \mathrm{d}k,$$

it follows from (41)′ that $\exp(-\beta H_I(h))$ may be expressed in the form

$$\exp(-\beta H_I(h)) = C \int \mathrm{d}K \exp\left(-\frac{K^2}{2\beta J} - \mathrm{i}K \,.\, (\sigma - \tilde{\sigma} - h_I)\right),$$

where K is a multidimensional vector and C a constant. Hence, by (42), $Z_\Lambda(h)$ takes the form

$$Z_\Lambda(h) = C \int \mathrm{d}K \exp\left(-\frac{K^2}{2\beta J} + \mathrm{i}K \,.\, h_I\right)$$
$$\times \langle \exp(-\mathrm{i}K \,.\, \sigma - BH(\Lambda_1, h_1)) \exp(\mathrm{i}K \,.\, \tilde{\sigma} - \beta H(\Lambda_2, h_2)) \rangle_{0,\Lambda}.$$

By (31), (35), and (40), the last exponential in this formula is identical to $\theta \exp(-\mathrm{i}K \,.\, \sigma - \beta H(\Lambda_1, h_2))$, and therefore

$$Z_\Lambda(h) = C \int \mathrm{d}K \exp\left(-\frac{K^2}{2\beta J} + \mathrm{i}K \,.\, h_I\right)$$
$$\langle \exp(-\mathrm{i}K \,.\, \sigma - \beta H(\Lambda_1, h_1)) \theta \exp(-\mathrm{i}K \,.\, \sigma - \beta H(\Lambda_1, h_2)) \rangle_{0,\Lambda}. \tag{43}$$

From this equation and the RP inequality (34), it follows that

$$Z_\Lambda(h) \equiv |Z_\Lambda(h)| \leqslant C \int \mathrm{d}K \exp\left(-\frac{K^2}{2\beta J}\right)$$
$$\times \langle \exp(-\mathrm{i}K\,.\,\sigma - \beta H(\Lambda_1, h_1))\theta \exp(-\mathrm{i}K\,.\,\sigma - \beta H(\Lambda_1, h_1))\rangle^{\frac{1}{2}}_{0,\Lambda}$$
$$\times \langle \exp(-\mathrm{i}K\,.\,\sigma - \beta H(\Lambda_1, h_2))\theta \exp(-\mathrm{i}K\,.\,\sigma - \beta H(\Lambda_1, h_2))\rangle^{\frac{1}{2}}_{0,\Lambda}.$$

Hence, by the Schwartz inequality $|\int fg\,\mathrm{d}K| \leqslant (\int |f|^2\,\mathrm{d}K)^{\frac{1}{2}}(\int |g|^2\,\mathrm{d}K)^{\frac{1}{2}}$,

$$Z_\Lambda(h) \leqslant \left[C \int \mathrm{d}K \exp\left(-\frac{K^2}{2\beta J}\right)\langle \exp(-\mathrm{i}K\,.\,\sigma - \beta H(\Lambda_1, h_1))\theta \right.$$
$$\left. \times \exp(-\mathrm{i}K\,.\,\sigma - \beta H(\Lambda_1, h_1))\rangle_{0,\Lambda}\right]^{\frac{1}{2}}$$
$$\times \left[C \int \mathrm{d}K \exp\left(-\frac{K^2}{2\beta J}\right)\langle \exp(-\mathrm{i}K\,.\,\sigma - \beta H(\Lambda_1, h_2))\theta \right.$$
$$\left. \times \exp(-\mathrm{i}K\,.\,\sigma - \beta H(\Lambda_1, h_2))\rangle_{0,\Lambda}\right]^{\frac{1}{2}}. \tag{44}$$

On the other hand, if h is replaced by h_j ($j = 1$ or 2) in the formula (43) then, by (35), the terms h_1, h_2, and h_I on the RHS of that formula have to be replaced by h_j, h_j, and 0, respectively, i.e.

$$Z_\Lambda(h_j) = C \int \mathrm{d}K \exp\left(-\frac{K^2}{2\beta J}\right)$$
$$\times \langle \exp(-\mathrm{i}K\,.\,\sigma - \beta H(\Lambda_1, h_j))\theta \exp(-\mathrm{i}K\,.\,\sigma - \beta H(\Lambda_1, h_j))\rangle_{0,\Lambda}.$$

It follows immediately from this equation and (44) that

$$Z_\Lambda(h) \leqslant (Z_\Lambda(h_1) Z_\Lambda(h_2))^{\frac{1}{2}},$$

which is the required inequality (36). This completes the proof of the order–disorder transition of the model.

5.3.4. *Generalizations of the FSS and Peierls arguments*

Let us now briefly discuss what is involved in generalizing the FSS and Peierls arguments to quantum systems. For definiteness, we consider cases where Σ is a quantum spin system on a simple cubic lattice X and utilize Griffiths's theorem, which tells us that Σ will exhibit low-temperature ordering if the finite version $\Sigma(\Lambda)$ of this system satisfies the long-range correlation condition (3). In order that we may utilize RP arguments for placings of the reflecting plane between any pair of neighbouring crystallographic planes of Λ, we shall again take this region to be a cube with periodic boundary conditions: as we shall presently see, the Peierls argument can also be adapted to these conditions.

In general, the canonical expectation functional for $\Sigma(\Lambda)$ is given by

$$\langle \ldots \rangle_\Lambda = \langle \ldots \exp(-\beta H(\Lambda))\rangle_{0,\Lambda} / \langle \exp(-\beta H(\Lambda))\rangle_{0,\Lambda} \tag{45}$$

as in (13), but now with

$$\langle \ldots \rangle_{0,\Lambda} = N(\Lambda)^{-1}\,\mathrm{Tr}_\Lambda(\ldots), \tag{46}$$

where $N(\Lambda)$ is the dimensionality of the Hilbert space $\mathscr{H}(\Lambda)$ of $\Sigma(\Lambda)$. As in the FSS argument, we consider the division of Λ into two halves, Λ_1 and Λ_2, that are mirror images of one another in a plane Π, containing no sites, the image of x ($\in \Lambda_1$) in Π being denoted by $\tilde{x}$. We define $\mathscr{A}_1$ and $\mathscr{A}_2$ to be the algebras generated by the spins in Λ_1, Λ_2, respectively and define a *reflection operator* θ to be an invertible mapping transformation of $\mathscr{A}_1$ onto $\mathscr{A}_2$ such that

$$\theta(\lambda A + \mu B) = \bar{\lambda}\theta A + \bar{\mu}\theta B \text{ for complex } \lambda, \mu, \tag{47}$$

$$\theta(AB) = (\theta A)(\theta B), \tag{48}$$

$$\theta I = I, \tag{49}$$

and

$$\mathrm{Tr}_{\Lambda_2}(\theta A) = \overline{\mathrm{Tr}_{\Lambda_1}(A)}. \tag{50}$$

By (46), this last condition ensures that the expectation functional $\langle \ldots \rangle_{0,\Lambda}$ has the RP property

$$\langle A\theta A\rangle_{0,\Lambda} \geqslant 0 \quad \text{for } A \text{ in } \mathscr{A}_1. \tag{51}$$

We shall assume that θ maps the spin $\sigma(x)$ at x into some element, $\tilde{\sigma}(\tilde{x})$, of the algebra of the spins at $\tilde{x}$, i.e.

$$\theta\sigma(x) = \tilde{\sigma}(\tilde{x}). \tag{52}$$

Hence, by (47) and (48), the mapping $\sigma(x) \to \tilde{\sigma}(\tilde{x})$ effectively determines θ, since

$$\theta \sum \lambda(j_1, \ldots j_k)\sigma_{j_1}(x^{(1)}) \ldots \sigma_{j_k}(x^{(k)}) = \sum \bar{\lambda}(j_1, \ldots j_k)\tilde{\sigma}_{j_1}(\tilde{x}^{(1)}) \ldots \tilde{\sigma}_{j_k}(\tilde{x}^{(k)}),$$

where $(\sigma_1(x), \sigma_2(x), \sigma_3(x))$ are the components of $\sigma(x)$. Furthermore, it follows from this formula that the conditions (47)–(50) are all fulfilled if the matrix elements of $\tilde{\sigma}(\tilde{x})$ are the complex conjugate of those of $\sigma(\tilde{x})$ in some representation. This evidently implies that there are various alternative choices of θ, corresponding to different representations. In concrete situations, the problem is to choose a reflection operator that may serve as the basis of an effective RP argument. Two examples of such reflection operators are given by the following forms for $\tilde{\sigma}$:

$$\tilde{\sigma}(\tilde{x}) = (\sigma_1(\tilde{x}), \sigma_2(\tilde{x}), -\sigma_3(\tilde{x})), \tag{53a}$$

and

$$\tilde{\sigma}(\tilde{x}) = -\sigma(\tilde{x}). \tag{53b}$$

In the first case, the matrix elements of $\tilde{\sigma}(\tilde{x})$ are the complex conjugates of those of $\sigma(\tilde{x})$ in the representation where $\sigma_1(\tilde{x})$ is diagonal and the matrices for the creation and annihilation operators $\sigma_2(x) \pm \mathrm{i}\sigma_3(x)$ are real: the second case is obtained from the first by a unitary transformation, corresponding to a rotation through 180° about Ox_3.

Note. One *cannot* take $\tilde{\sigma}(\tilde{x})$ to be $\sigma(\tilde{x})$, as in the classical case, since it follows from eqns (47), (48), and the commutation rule

$$[\sigma_1(x), \sigma_2(x)] = \mathrm{i}\sigma_3(x),$$

that

$$[\tilde{\sigma}_1(\tilde{x}), \tilde{\sigma}_2(\tilde{x})] = -\mathrm{i}\tilde{\sigma}_3(x);$$

and therefore, if $\tilde{\sigma}(\tilde{x})$ were $\sigma(\tilde{x})$, the commutation relation

$$[\sigma_1(\tilde{x}), \sigma_2(\tilde{x})] = +\mathrm{i}\sigma_3(\tilde{x})$$

would be violated.

Systems with continuous symmetry: *infra-red bounds*. In order to discuss the generalization of the FSS argument to quantum systems we first recall that, for the classical Heisenberg model, the interaction Hamiltonian between the systems in Λ_1 and Λ_2 took the form (cf. (41))

$$H_I(h) = \tfrac{1}{2}J \sum_{x \in \Pi_1} (\sigma(\tilde{x}) - \tilde{\sigma}(\tilde{x}) - h(x, \tilde{x}))^2, \text{ where } \tilde{\sigma}(\tilde{x}) = \theta\sigma(x), \tag{54}$$

with $J > 0$. Dyson, Lieb, and Simon (DLS) (1978) have generalized the FSS argument, to quantum spin systems with continuous symmetries for which the coupling between the parts in Λ_1 and Λ_2 reduced to this form, for some reflection operator θ. The essential mathematical tool that enabled them to carry the argument through for non-commutative algebras was the Trotter product formula

$$\exp\left(\sum_1^n A_n\right) = \lim_{N \to \infty}(\exp(A_1/N) \ldots \exp(A_n/N))^N.$$

Using this formula, DLS were able to adapt the FSS argument in such a way as to obtain the required infra-red bounds for a class of models for which the coupling between the parts of $\Sigma(\Lambda)$ in Λ_1 and Λ_2 took the form (54) for some reflection operator θ. Here, we shall specify the particular forms of θ for two of the models for which DLS proved low temperature ordering.

X–Y model. Here the interactions are between neighbouring Pauli spins and take the form

$$-J(\sigma_1(x)\sigma_1(x') + \sigma_2(x)\sigma_2(x')),$$

or, adding a constant term,

$$\tfrac{1}{2}J(\sigma(x)-\sigma(x'))_2^2 \equiv \tfrac{1}{2}J((\sigma_1(x)-\sigma_1(x'))^2+(\sigma_2(x)-\sigma_2(x'))^2), \text{ with } J>0.$$

In the presence of a perturbing field $h \equiv \{h(x,x')\}$, this interaction is changed to

$$\tfrac{1}{2}J(\sigma(x)-\sigma(x')-h(x,x'))_2^2,$$

and so the coupling between Λ_1 and Λ_2 takes the form

$$\tfrac{1}{2}J\sum_{x\in\Pi_1}(\sigma(x)-\sigma(\tilde{x})-h(x,\tilde{x}))_2^2.$$

Hence, choosing θ to be the reflection operator corresponding to the form (53*a*) for $\tilde{\sigma}$, this coupling reduces to

$$\tfrac{1}{2}J\sum_{x\in\Pi_1}(\sigma(x)-\tilde{\sigma}(\tilde{x})-h(x,\tilde{x}))_2^2 \quad (\theta\sigma(x)\equiv\tilde{\sigma}(\tilde{x})),$$

which is of the required form (cf. (54)).

The Heisenberg antiferromagnet. Here the interactions are between neighbouring spins and take the form $J\sigma(x)\,.\,\sigma(x')$ or equivalently

$$\tfrac{1}{2}J(\sigma(x)+\sigma(x'))^2, \quad \text{with} \quad J>0.$$

In the presence of a perturbing field $h=\{h(x,x')\}$, this is changed to

$$\tfrac{1}{2}J(\sigma(x)+\sigma(x')-h(x,x'))^2,$$

and so the coupling between Λ_1 and Λ_2 takes the form

$$\tfrac{1}{2}J\sum_{x\in\Pi_1}(\sigma(x)+\sigma(\tilde{x})-h(x,\tilde{x}))^2.$$

Hence, choosing θ to be the reflection operator corresponding to the form (53*b*) for $\tilde{\sigma}$, this is equivalent to

$$\tfrac{1}{2}J\sum_{x\in\Pi_1}(\sigma(x)-\tilde{\sigma}(\tilde{x})-h(x,\tilde{x}))^2 \quad (\theta\sigma(x)=\tilde{\sigma}(x)),$$

which is of the required form (54), leading to the conclusion that the free energy $F_\Lambda(h)$ is minimized when $h=0$, as in the case of the classical Heisenberg ferromagnet. However, the counterpart of the condition (22) for that model is not so simple here, and the proof that this ensues from the minimization of F_Λ at $h=0$ has been achieved only for the cases when either the spin $\geqslant 1$ or when the dimensionality $\geqslant 4$.

Note on Heisenberg ferromagnet. For this model, as for its classical version, the coupling between Λ_1 and Λ_2 is of the form

$$\tfrac{1}{2}J\sum_{x\in\Pi_1}(\sigma(x)-\sigma(\tilde{x})-h(x,\tilde{x}))^2.$$

Here, however, as noted above, there is no reflection operator θ for which $\theta\sigma(x) = \tilde{\sigma}(x)$, and consequently one cannot express this interaction in a form amenable to the RP arguments. It is for this reason that the present techniques have not led to a proof of ordering in the Heisenberg ferromagnet.

Systems with discrete symmetries: *generalized Peierls argument*. The original form of the Peierls argument for the Ising ferromagnet was based on boundary conditions of the positive (or negative) type (cf. §3.5.1). However, one can adapt the same argument to periodic boundary conditions for the following reasons. By Griffiths's theorem, as represented by eqn (3), a *sufficient* condition for ordering in the infinite system Σ is that the correlation between spins at points x and y of the finite one, $\Sigma(\Lambda)$, with periodic conditions, tends to a non-zero limit when first Λ and then $|x-y|$ tends to infinity. Hence, to prove ordering, it suffices to show that the probability $P_\Lambda(x,y)$ that the spins at x and y take opposite values is less than some constant $b < \frac{1}{2}$ for sufficiently large $|x-y|$. Now, for any configuration in which the spins at x and y have opposite signs, at least one of these points must be surrounded by a Peierls contour, separating positive and negative spins. $P_\Lambda(x,y)$ may therefore be estimated in terms of Peierls contour probabilities, as in §3.5.1, and these lead to the required low-temperature result that $P_\Lambda(x,y) < b < \frac{1}{2}$ for $|x-y|$ sufficiently large. Thus, the argument can be carried through perfectly well for periodic boundary conditions.

In order to cast this argument into a form that can be generalized to other systems, we note that $P_\Lambda(x,y)$ is given by the formula

$$P_\Lambda(x,y) = \langle p^{(+)}(x)p^{(-)}(y) + p^{(-)}(x)p^{(+)}(y)\rangle_\Lambda, \tag{55}$$

where $p^{(\pm)}(x)$ are the projection operators corresponding to positive and negative spin values at x. Likewise, the probability that a contour γ is a boundary, separating regions of positive and negative spins, is given by

$$P_\Lambda(\gamma) = \left\langle \prod_{x\in\gamma_{\text{in}}} \prod_{y\in\gamma_{\text{ex}}} (p^{(+)}(x)p^{(-)}(y) + p^{(-)}(x)p^{(+)}(y)) \right\rangle_\Lambda, \tag{56}$$

where γ_{in} and γ_{ex} consist of the sites nearest to γ lying respectively inside and outside that contour. Furthermore, as $p^{(+)}(x) + p^{(-)}(x) = I$, one can use these last two equations to express $P_\Lambda(x,y)$ in terms of Peierls contour probabilities $P_\Lambda(\gamma)$, which in turn are correlation functions for certain projection operators. Now this is also true for other spin systems, including bona fide quantum mechanical ones, though in general the correlation functions for the Peierls contour probabilities are much more complicated than in the Ising case. It is here that RP arguments can sometimes be used to provide manageable estimates for these probabilities. Let us now indicate how this has been achieved by J. Fröhlich, Israel, Lieb, and Simon (FILS) (1978, 1980).

The method developed by FILS is applicable to systems for which the coupling between the parts of the system in the regions Λ_1 and Λ_2, into which Π divides Λ, is of the form

$$H_I = -\sum_r C_r \theta C_r \tag{57}$$

for some reflection operator θ and some observables $\{C_r\}$ in Λ_I. For such systems, FILS have deduced from the properties (47)–(50) of θ and the RP of $\langle \ldots \rangle_{0,\Lambda}$ that the canonical expectation functional $\langle \ldots \rangle_\Lambda$ is also RP, i.e.

$$\langle A\theta A \rangle_\Lambda \geqslant 0, \tag{58}$$

and therefore that the following Schwartz inequality holds:

$$|\langle A\theta B \rangle_\Lambda| \leqslant \langle A\theta A \rangle_\Lambda^{\frac{1}{2}} \langle B\theta B \rangle_\Lambda^{\frac{1}{2}}. \tag{59}$$

We remark that the class of systems satisfying the condition (57) is a large one, including all the models mentioned in §5.3.2 among others (cf. FILS). For example, in the case of the Ising ferromagnet, H_I takes the form $-J \sum_{x\in\Pi_1} \sigma_1(x)\theta\sigma_1(x)$ with $\theta\sigma(x) = \tilde{\sigma}(\tilde{x})$, as given by (53*a*); and for the anisotropic Heisenberg antiferromagnet, $H_I = -J \sum_{x\in\Pi_1} (\alpha\sigma_1(x)\theta\sigma_1(x) + \alpha\sigma_2(x)\theta\sigma_2(x) + \sigma_3(x)\theta\sigma_3(x))$, with $\theta\sigma(x) = \tilde{\sigma}(\tilde{x})$ given by (53*b*).

For systems satisfying (57), FILS employed the Schwartz inequalities (59) in such a way as to reduce the problem of estimating the various Peierls contour probabilities to the manageable one of obtaining an upper bound for the probability of a single 'universal contour' along the boundary of Λ. In this way, they obtained upper bounds for these probabilities that sufficed to prove low-temperature ordering in a variety of models with discrete symmetries, e.g. the two dimensional anisotropic Heisenberg ferromagnet, with spin $\geqslant 1$, and certain Coulomb systems.

5.4. Critical phenomena

Critical points of order–disorder transitions are characterized by thermodynamical singularities and by infinite-range correlations in the fluctuations of the order field (cf. eqn (6)). This infinite-range correlation property is especially significant because, as observed by Jona-Lasinio (1975) and Sinai (1976, 1982), it implies that the Central Limit Theorem, which is basic to Statistical Mechanics,† is inapplicable at the critical point. In order to explain this observation, we recall that if $\xi_1, \ldots, \xi_n$ are n independent random variables, each with zero mean and dispersion σ, then the Central Limit Theorem tells us that, in the limit $n \to \infty$, the variable $\xi = n^{-\frac{1}{2}}(\xi_1 + \ldots + \xi_n)$ is distributed with Gaussian probability density

$$P(\xi) = (2\pi\sigma^2)^{-\frac{1}{2}} \exp(-\xi^2/2\sigma^2).$$

† Cf. Khinchin (1949).

This, of course, depends on the fact that

$$\lim_{n\to\infty} n^{-1}\langle(\xi_1 + \ldots + \xi_n)^2\rangle = \sigma^2$$

in this case. The Central Limit Theorem is also applicable in more general circumstances in Statistical Mechanics, even when there are correlations between the variables $\xi_1, \ldots, \xi_n$. Thus, the standard situation, as applied to translationally invariant states ρ of a lattice system Σ is that the observable

$$\hat{\eta}_\Lambda = |\Lambda|^{-\frac{1}{2}} \sum_{x\in\Lambda} \hat{\eta}(x)$$

is distributed with the Gaussian probability density in the limit $\Lambda \to \infty$, with a dispersion, σ, given by

$$\sigma^2 = \lim_{\Lambda\uparrow} |\Lambda|^{-1}\rho\Big(\Big(\sum_{x\in\Lambda} \hat{\eta}(x)\Big)^2\Big).$$

However, in the case of the critical state ρ_c, this limit is not finite! For, by the correlation property (6) and the translational invariance of ρ_c

$$\lim_{\Lambda\uparrow} |\Lambda|^{-1}\rho_c\Big(\Big(\sum_{x\in\Lambda} \hat{\eta}(x)\Big)^2\Big) = \sum_{x\in X} \rho_c(\hat{\eta}(x)\,.\,\hat{\eta}(0)) = \infty, \tag{60}$$

and consequently the Central Limit Theorem is inapplicable to ρ_c. This evidently means that the bulk properties of critical states should be characterized by some kind of *non-central limiting behaviour*. The theory of critical phenomena must therefore be based on a structure that is not standard in statistical mechanics, and our objective in this section is to describe developments made towards the determination of that structure.

We shall start, in §5.4.1, by discussing the background of ideas (mainly phenomenological) leading to the view propounded by Kadanoff (1966) and Wilson (1971) that the bulk properties of a system at its critical point are essentially scale invariant. This leads naturally to a description of critical phenomena in terms of renormalizations, i.e. scale transformations. In §5.4.2, we shall present a quantum mechanical adaptation of Sinai's probabilistic formulation of the renormalization theory which reduces the theory of critical phenomena to a problem of the stability of certain gross properties of scale invariant states (a problem in fixed point analysis). In §5.4.3, we shall present a fairly detailed account of the treatment, by Bleher and Sinai (1973, 1975), of a special class of models (the hierarchical ones of Dyson (1969)) within the terms of Sinai's general framework. This serves to evaluate the indices, representing the critical singularities of the models, and to exhibit their lack of dependence on various microscopic details. Finally, in §5.4.4, we shall briefly discuss the relationship between Sinai's rigorous method, whose range of proven applicability is still very limited, and the more heuristic approach of Wilson and Fisher (1972), which has provided accurate calculations of critical indices for a wide class of models. We shall also briefly discuss Griffith's (1981) observation that there is still a gap between the microscopic and large scale descriptions of many-particle systems in the renormalization theory.

5.4.1. *Phenomenological background*

Landau's theory. Let us start by briefly recalling Landau's phenomenological treatment of critical singularities.† This was based on the assumptions that the Helmholtz potential $\psi(\eta, T)$, expressed as a function of the order parameter and temperature, could be written as a power series in η, with coefficients that varied smoothly with T; and that the system had an internal symmetry corresponding to reversals $(\eta \to -\eta)$ or rotations $(\eta \to R\eta;\ |R\eta| = |\eta|)$ of the order parameter, which ensured that $\psi(\eta, T)$ was a function of η^2 and T only. Under these assumptions, ψ could be expressed in the form

$$\psi(\eta, T) = A_0(T) + A_1(T)\,|\eta|^2 + A_2(T)\,|\eta|^4 + \ldots, \tag{61}$$

the higher-order terms being negligible for the critical region, where $|\eta|$ is small. Landau's essential idea was that a phase transition of the second kind corresponded to a change of $A_1(T)$ from negative to positive values as T increased through T_c. For, assuming‡ that $A_1(T)$ behaves in this way and that $A_2(T)$ is positive, it follows from (61) that at equilibrium, where $\psi(\eta, T)$ is minimized w.r.t. η,

$$\eta = 0 \quad \text{for } T \geqslant T_c$$

and

$$|\eta| \simeq (-A_1(T)/2A_2(T))^{\frac{1}{2}} \quad \text{for} \quad T < T_c.$$

Correspondingly, the equilibrium free energy is given by

$$\phi(T) = A_0(T) \quad \text{for} \quad T \geqslant T_c$$

and

$$\phi(T) \simeq A_0(T) - A_1(T)^2/4A_2(T) \quad \text{for} \quad T < T_c.$$

Assuming further that the first derivative of $A_1(T)$ does not vanish at T_c, these formulae for η and ϕ simplify to the following forms in the critical region.

$$|\eta| = \begin{cases} 0 \quad \text{for} \quad T > T_c \\ \text{const.}\,|T - T_c|^{\frac{1}{2}} \quad \text{for} \quad T < T_c \end{cases} \tag{62}$$

† See Landau and Lifshitz (1959), §135.

‡ Strictly speaking, this assumption is untenable, as it violates the requirement that ψ is convex in η. However, the situation may be remedied by redefining the free energy to be the modification of ψ as given by Maxwell's construction (cf. Lebowitz and Penrose 1966).

and

$$\phi(T) = \begin{cases} A_0(T) & \text{for} \quad T > T_c \\ A_0(T) - \text{const.}\,(T - T_c)^2 & \text{for} \quad T < T_c. \end{cases} \tag{63}$$

Since the specific heat, $K(T)$, is $-T\,\mathrm{d}^2\phi(T)/\mathrm{d}^2T$, it follows from (63) that this has a finite discontinuity at T_c. Thus, according to Landau's theory, there is a *universal law* that the order parameter varies as $|T_c - T|^{\frac{1}{2}}$ as T approaches T_c from below and that the specific heat has a finite discontinuity at the critical point. This, in fact, agrees precisely with the Van der Waals theory of the liquid–gas transition. In general, the critical behaviour of the Landau–Van der Waals theories is realized by models† with weak, long-range interactions of the form $\alpha^d V(\alpha(x - x'))$, as formulated in the limit $\alpha \to 0$, where d is the dimensionality and $V(x)$ is absolutely integrable (e.g. $V(x) = \mathrm{e}^{-|x|}$): these are *mean field theoretic* models in which the effective potential in which each particle moves is of the self-consistent, Hartree type, and so carries no fluctuations.

In fact, the validity of Landau's theory appears to be restricted to mean field theoretic models, since it conflicts with the extensive results‡ provided by exact treatments of solvable models, numerical computations and experimental measurements. Most significantly, these results all signify that the free energy density, $\phi(T)$, and thus also the specific heat, should have an essential singularity at T_c – logarithmic in the case of Onsager's solution of the two-dimensional Ising model – whereas according to Landau's theory, $\phi(T)$ is piecewise analytic (cf. (63)). From this, one can infer that Landau's basic assumption of the analyticity of the Helmholtz potential, $\psi(\eta, T)$, is generally invalid, since it is this that implies the piecewise analytic property of $\phi(T)$. Presumably, this defect in Landau's theory reflects the fact that it is essentially mean field theoretic and so fails to represent the giant fluctuations that characterize critical points. Progress beyond Landau's theory therefore requires a treatment of critical phenomena that takes proper account of these fluctuations.

Critical singularities. Before considering general theoretical questions any further, we shall now briefly describe the basic facts concerning critical singularities that have emerged from exact solutions of models, computations and experiments (cf. Fisher 1967). Thus, taking the independent thermodynamical variables to be the external field y, conjugate to the order, $\hat{\eta}$, and the reduced temperature

$$t = (T - T_c)/T_c, \tag{64}$$

we denote the Gibbs potential by $\phi(y, t)$ and formulate the specific heat,

† See Lebowitz and Penrose (1966), Lieb (1966), and Emch (1967*a*).
‡ See the review article by Fisher (1967*b*).

$K(t)$, at $y=0$, the spontaneous polarization (or order) $\eta(t)$, below the critical point, the generalized susceptibility, $\chi(t)$, above it and the field-dependent polarization, $\eta_c(y)$, at $T=T_c$ by the following equations:

$$K(T) = -\frac{T}{T_c^2}\frac{\partial^2\phi(0,t)}{\partial t^2} \simeq -T_c^{-1}\frac{\partial^2\phi(0,t)}{\partial t^2} \tag{65}$$

$$\eta(t) = \lim_{y\to 0} -\frac{\partial\phi(0,t)}{\partial y}, \quad \text{for} \quad t<0, \tag{66}$$

the limit being taken along a fixed direction for y,

$$\chi(t) = -\left(\frac{\partial^2\phi(y,t)}{\partial y^2}\right)_{y=0}, \quad \text{for} \quad t>0 \tag{67}$$

and

$$\eta_c(y) = -\frac{\partial\phi(y,0)}{\partial y}. \tag{68}$$

The theoretical and experimental investigations on the behaviour of these functions near the critical point reveal that they are all governed by simple power laws, possibly complicated by logarithms, i.e.

$$K(t) \sim |t|^{-\alpha} \quad (\text{or } |t|^{-\alpha}\ln|t|), \quad \text{with} \quad \alpha \geqslant 0, \tag{69}$$

$$\chi(t) \sim t^{-\gamma} \quad (\text{or } t^{-\gamma}\ln t), \quad \text{with} \quad \gamma>0\,(t>0), \tag{70}$$

$$|\eta(t)| \sim |t|^{\beta}, \quad \text{with} \quad \beta>0\,(t<0), \tag{71}$$

and

$$|\eta_c(y)| \sim |y|^{\delta}, \quad \text{with} \quad \delta>0. \tag{72}$$

The constants α, β, γ, and δ appearing in these formulae are termed *critical indices*.

Remarkably, the values of these indices are found to depend only on very general qualitative features of the system, such as symmetry, dimensionality, and interaction range (finite or infinite).† Thus, for example, the critical indices of Ising spin systems with short-range ferromagnetic interactions depend on the dimensionality of the lattice but not on the detailed form of either the interactions (e.g. whether confined to nearest neighbours) or the lattice (e.g. whether cubic or hexagonal). Moreover, the critical indices for this class of spin systems are the same as those for a liquid–gas transition in a fluid with short-range interactions, the counterparts to spontaneous magnetization and susceptibility being the difference between liquid and gaseous densities and the isothermal compressibility, respectively. Similarly, the critical indices of Heisenberg-type spin systems with

† cf. Fisher (1967*b*).

rotationally symmetrical short-range interactions depend on the dimensionality but not on the details of the interactions. These indices, however, are different from those of the Ising systems, this difference presumably stemming from that between continuous and discrete internal symmetries.

Thus, although the classification of phase transitions of the second kind according to the values of the critical indices is governed only by very general qualitative features of the systems concerned, it is nevertheless more complex than that given by the Landau theory, which predicted a universal law for these indices (cf. (62) and (63)).

Scaling laws. The power laws governing the critical singularities have been brought together by Widom's (1965) observation that, according to a wealth of theoretical and experimental evidence, the thermodynamics of the critical region follows simple *scaling laws*. These may be summarized by the statement that the value of the Gibbs potential $\phi(y,t)$ is invariant under transformations $y \to \lambda^a y$, $t \to \lambda^b t$ and $\phi \to \lambda^{-1}\phi$ for certain indices a and b and arbitrary positive λ, i.e.

$$\lambda\phi(y,t) \equiv \phi(\lambda^a y, \lambda^b t) \quad \text{for positive } \lambda. \tag{73}$$

This scaling property implies simple interrelationships between the critical indices α, β, γ, and δ. For, by (73), $\phi(0,t)$ and $\phi(y,0)$ satisfy the homogeneity conditions

$$\phi(0, rt) = r^{1/b}\phi(0,t) \quad (\text{with } r = \lambda^{1/b})$$

and

$$\phi(ry, 0) = r^{1/a}\phi(y,0) \quad (\text{with } r = \lambda^{1/a}).$$

Hence, $\phi(0,t)$ and $\phi(y,0)$ vary as $t^{1/b}$ and $|y|^{1/a}$, respectively, and therefore, by (65), (68), (69), and (72)

$$\alpha = 2 - b^{-1} \quad \text{and} \quad \delta = a^{-1} - 1. \tag{74}$$

Furthermore, by eqns (66), (67), and (73),

$$\lambda\eta(t) = \lambda^a \eta(\lambda^b t)$$

and

$$\lambda\chi(t) = \lambda^{2a}\chi(t),$$

i.e.

$$\eta(rt) = r^{(1-a)/b}\eta(t) \quad (\text{with } r = \lambda^{1/b})$$

and

$$\chi(rt) = r^{(1-2a)/b}\chi(t) \quad (\text{with } r = \lambda^{1/b}).$$

Hence, $\eta(t) \sim |t|^{(1-a)/b}$ and $\chi(t) \sim t^{(1-2a)/b}$, and consequently, by (70) and (71)

$$\beta = (1-a)/b \quad \text{and} \quad \gamma = (2a-1)/b.$$

On combining these equations with (74), we see that the critical indices are interrelated according to the formulae

$$\left.\begin{aligned} \alpha + \beta + \gamma &= 2 \\ \text{and} \qquad \delta^{-1} &= 1 + \gamma/\beta, \end{aligned}\right\} \tag{75}$$

These formulae tell us that *there are only two independent critical indices*.

Connection between scaling and critical fluctuations: Kadanoff's argument. Suppose that Σ has a unique equilibrium state $\rho_{y,t}$ for the values (y, t) of the thermodynamical variables. Then the correlation function, governing the fluctuations in the order field for this state is

$$C_{y,t}(x) = \rho_{y,t}(\hat{\eta}(x) \,.\, \hat{\eta}(0)) - \rho_{y,t}(\hat{\eta}(x)) \,.\, \rho_{y,t}(\hat{\eta}(0)). \tag{76}$$

Assuming that $C_{y,t}(x)$ falls off exponentially with $|x|$ at large distances, which it does in 'good' cases,† provided that (y, t) is not the critical point $(0, 0)$, one may define a correlation length $\xi(y, t)$ by the condition that $C_{y,t}(x) \sim \exp(-|x|/\xi(y, t))$ or, more precisely,

$$\xi(y, t) = \sup\{l \mid C_{y,t}(x) \exp(|x|/l) \to 0 \quad \text{as } |x| \to \infty\}. \tag{77}$$

We shall assume this formula also at the critical point, $y = 0$, $t = 0$, where, in view of (6), it leads to an infinite value for ξ. Correspondingly we shall assume that, by continuity, $\xi(y, t) \to \infty$ as $(y, t) \to 0$, i.e. that ξ is 'very large' in the critical region.

Kadanoff (1966) made an important connection between this long-range correlation property and the scaling laws by the following heuristic argument. Suppose, for definiteness, that Σ is a system of spins with short-range interactions on a d-dimensional cubic lattice, X, with unit spacing. Then we may resolve X into cubes $\{\Delta_L(x)\}$ of side L, an integer, centred at the sites $\{Lx\}$ of the lattice X_L obtained by dilating X by the factor L. Σ may therefore be regarded as a system of 'blocks', comprised by the subsystems of spins in these cubes. Since the block formed by the spins in $\Delta_L(x)$ is centred at Lx, it follows that we may map Σ into a system Σ_L of blocks, centred at the sites of the original lattice X, by rescaling length by a factor L^{-1}, without changing the interaction energies. The free energy density $\phi_L(y, t)$ of Σ_L is then L^{-d} times that of Σ, i.e.

$$\phi_L(y, t) \equiv L^{-d}\phi(y, t). \tag{78}$$

† See, for example, Lebowitz and Penrose (1968).

Kadanoff supplemented this identity with a second relationship between ϕ_L and ϕ, designed exclusively for the critical region, where $\xi \gg L$ or, equivalently, where the correlation length, ξ/L, of Σ_L is much greater than the lattice spacing. This was based on the observation that the block attached to x carried a resultant spin, $\hat{\eta}_L(x)$, and so Σ_L could be regarded as a system of block spins, $\hat{\eta}_L(x)$, which interacted via forces with the same short range and symmetry properties as those of Σ. Thus, invoking the hypothesis that the critical behaviour of a system is determined by its symmetry, dimensionality, and interaction range, Kadanoff assumed that the only respect in which the critical properties of Σ_L could differ from those of Σ corresponded to a change in the effective values of the thermodynamical control variables from (y, t) to (y_L, t_L), i.e.

$$\phi_L(y, t) = \phi(y_L, t_L). \tag{79}$$

We denote by $\tau(L)$ the transformation from (y, t) to (y_L, t_L), i.e.

$$(y_L, t_L) = \tau(L)(y, t). \tag{80}$$

Since $\tau(L)$ corresponds to a rescaling of length by the factor L, it follows that this transformation possesses the *semigroup* property

$$\tau(L_1)\tau(L_2) = \tau(L_1 L_2). \tag{81}$$

This formula is satisfied if

$$\tau(L)(y, t) = (L^p y, L^q t)$$

for some indices p and q, which is the solution assumed by Kadanoff. Taken together with eqns (79) and (80), it implies that

$$\phi_L(y, t) = \phi(L^p y, L^q t),$$

and therefore, by (78), that

$$\phi(L^p y, L^q t) = L^{-d}\phi(y, t). \tag{82}$$

This is identical to Widom's scaling relation (72), with $p = \mathrm{d}a$ and $q = \mathrm{d}b$.

Kadanoff's derivation of the scaling laws is, of course, based on a phenomenological argument that provides no means of relating the indices p and q, or equivalently the critical indices, to the microscopic properties of the system. Wilson (1971) subsequently devised a microscopic version of this argument, based on the idea that the bulk properties of critical states are invariant with respect to scale transformations, usually called renormalizations, that are analogous to $\{\tau(L)\}$. The renormalization theory has provided a new conception of critical phenomena as well as successful computations of critical indices. Sinai (1982, Ch. 3) has recast it into a precise mathematical form, within the terms of probability theory. Here we shall present a quantum mechanical version of Sinai's formulation.

5.4.2. Renormalization theory and non-central limits

Characteristic functions. We again assume that Σ is a system of particles, or spins, on a d-dimensional lattice, X, with an order field, $\hat{\eta}(x) = (\hat{\eta}_1(x), \ldots, \hat{\eta}_n(x))$, whose components do not, in general, commute. We shall always take it that $\hat{\eta}(x)$ is bounded. The non-commutativity of the components of $\hat{\eta}(x)$ evidently precludes the possibility of representing the statistical properties of this field by a classical probability distribution. However, as the local observables given by functions of this field are generated by unitary operators of the form $\exp i(\sum_{j=1}^{k} \xi_j \cdot \hat{\eta}(x_j))$, with $(x_1, \ldots, x_k)$ an arbitrary set of lattice points and the components $(\xi_{j,1}, \ldots, \xi_{j,n})$ of ξ_j c-numbers, it follows that the statistical properties of $\hat{\eta}$ for a state ρ are governed by the *characteristic function*

$$\mu(\xi_1, \ldots, \xi_k) = \rho\left(\exp i \sum_{j=1}^{k} (\xi_j \cdot \hat{\eta}(x_j))\right). \tag{83}$$

Thus, μ is the quantum analogue of a classical characteristic function, defined as the Fourier transform of a probability. The formula (83) is equivalent to a definition of μ as a function of a classical field $\xi(x)$ $(= (\xi_1(x), \ldots, \xi_n(x)))$, which vanishes at all but a finite set of lattice sites, i.e.

$$\mu(\xi) = \rho\left(\exp i\left(\sum_x \xi(x) \cdot \hat{\eta}(x)\right)\right). \tag{83$'$}$$

Block variables and renormalization. We may formulate the statistical properties of the block variables of Kadanoff's argument in a similar way. For this, we again resolve the lattice X into cubes $\{\Delta_L(x)\}$ of side L, centred at the sites $\{Lx\}$ of the lattice X_L obtained by dilating X by the factor L. To be precise, we always take L to be even and define

$$\Delta_L(x) = \left\{Lx + y \mid y = (y_1, \ldots, y_d); y_j = \frac{L}{2}, -\frac{L}{2} + 1, \ldots, \frac{L}{2} \right.$$
$$\left. \text{for} \quad j = 1, \ldots, d\right\}. \tag{84}$$

We define the block variables, $\hat{\eta}_L(x)$, for the cubes, $\Delta_L(x)$, by the formula

$$\hat{\eta}_L(x) = (\hat{\eta}_{L,1}(x), \ldots, \hat{\eta}_{L,n}(x)) = L^{-md/2} \sum_{x' \in \Delta_L(x)} \hat{\eta}(x'), \tag{85}$$

where the index m is to be chosen so that the dispersion in $\hat{\eta}_L(x)$, at the critical point, tends to a finite non-zero limit as $L \to \infty$. It is easy to put bounds on the values that m could take. For on the one hand, the failure of the central limit theorem at the critical point arises because the dispersion in $\sum_{x \in \Lambda} \hat{\eta}(x)$ increases faster than $|\Lambda|^{\frac{1}{2}}$ there as $\Lambda \to \infty$ (cf. (60)); while, on the other hand, the boundedness of $\hat{\eta}(x)$ implies that this dispersion cannot increase faster than $|\Lambda|$. Hence, by (85), if the dispersion in $\hat{\eta}_L(x)$ is to tend to a finite non-zero limit as $L \to \infty$, then

$$1 < m \leqslant 2. \tag{86}$$

We remark that this condition implies that the block variables $\hat{\eta}_L(x)$ behave *classically*, in the sense that their components $(\hat{\eta}_{L,1}(x), \ldots, \hat{\eta}_{L,n}(x))$ intercommute, in the limit $L \to \infty$. For, by (85) and the commutativity of observables at different sites,

$$\|[\hat{\eta}_{L,j}(x), \hat{\eta}_{L,k}(x)]\| = 0(L^{d(1-m)}) \to 0 \quad \text{as} \quad L \to \infty, \quad \text{since } m > 1.$$

For finite L, the statistics of the field, $\hat{\eta}_L(x)$, of the block variables is governed by the characteristic function μ_L, obtained by replacing $\hat{\eta}$ by $\hat{\eta}_L$ in (83) or (83′), i.e.

$$\mu_L(\xi_1, \ldots, \xi_k) = \rho\left(\exp \mathrm{i} \sum_{j=1}^{k} \xi_j \cdot \hat{\eta}_L(x_j)\right), \tag{87}$$

or equivalently

$$\mu_L(\xi) = \rho\left(\exp \mathrm{i} \sum_{x} \xi(x) \cdot \hat{\eta}_L(x)\right). \tag{87$'$}$$

To express μ_L in terms of the characteristic function, μ, for the order field $\hat{\eta}(x)$, we note that, by (85) and (87)′,

$$\begin{aligned}\mu_L(\xi) &= \rho\left(\exp \mathrm{i} \sum_{x} \sum_{x' \in \Delta_L(x)} L^{-md/2} \xi(x) \cdot \hat{\eta}(x')\right)\\ &= \rho\left(\exp \mathrm{i} \sum_{x' \in X} \xi_L(x') \cdot \hat{\eta}(x')\right),\end{aligned} \tag{88}$$

where

$$\xi_L(x') = L^{-md/2}\xi(x) \quad \text{for} \quad x' \in \Delta_L(x). \tag{89}$$

Since the cells, $\Delta_L(x)$, cover the lattice and do not overlap, this last equation defines ξ_L at all sites of X. Hence, by (83)′ and (88),

$$\mu_L(\xi) = \mu(\xi_L). \tag{90}$$

Denoting by $\alpha(L)$ the transformation from ξ to ξ_L defined by (89), i.e.

$$(\alpha(L)\xi)(x') = \xi_L(x') = L^{-md/2}\xi(x) \quad \text{for} \quad x' \in \Delta_L(x), \tag{91}$$

it follows easily from (84) that $\alpha(L)$ has the semigroup property

$$\alpha(L_1)\alpha(L_2) = \alpha(L_1 L_2). \tag{92}$$

Correspondingly, by (90)–(92), the transformation $R(L)$ carrying μ to μ_L is given by

$$(R(L)\mu)(\xi) \equiv \mu_L(\xi) \equiv \mu(\alpha(L)\xi) \tag{93}$$

and possesses the analogous semigroup property

$$R(L_1)R(L_2) = R(L_1 L_2). \tag{94}$$

The transformations, $R(L)$, evidently correspond to length dilations $x \to Lx$ together with a 'collectivization' of the order field, $\hat{\eta} \to \hat{\eta}_L$. They are generally referred to as renormalizations, and we shall refer to $\{R(L)\}$ as the *renormalization semigroup* (RS). Note that this is not a group† because $R(L)$, like the transformations $\tau(L)$ of Kadanoff's argument, is not generally invertible. For classical systems, the RS corresponds to transformations of the probability distribution given by the Fourier transform of μ.

Critical states as fixed points of the RS. Following Sinai (1982, Ch. 3), we now aim to characterize critical states by non-central limiting behaviour. Thus we seek a limit distribution governing the block variables, $\hat{\eta}_L(x)$, in the critical state ρ_c, as $L \to \infty$. Here we anticipate that this distribution should be *classical* in view of the observation, following eqn (85), that the components of $\hat{\eta}_L(x)$ intercommute in the limit $L \to \infty$. We therefore look for a reduction of the statistics of the variables, $\hat{\eta}_L(x)$, in this limit, to those of classical ones, $\eta^{(cl)}(x)$, distributed with probability, P, as expressed by the condition

$$\lim_{L\to\infty} \rho_c(f(\hat{\eta}_L(x_1), \ldots, \hat{\eta}_L(x_k))) = P(f(\eta^{(cl)}(x_1), \ldots, \eta^{(cl)}(x_k))), \tag{95}$$

where $P(\ldots)$ denotes the expectation value for P and f is an arbitrary function of the form‡

$$f(\hat{\eta}_L(x_1), \ldots, \hat{\eta}_L(x_k)) \begin{cases} \sum \phi(\xi_1, \ldots, \xi_k) \exp i\left(\sum_1^k \xi_j \cdot \hat{\eta}_L(x_j)\right) \\ \text{or} \\ \int \phi(\xi_1, \ldots, \xi_k) \exp i\left(\sum_1^k \xi_j \cdot \hat{\eta}_L(x_j)\right). \end{cases} \tag{96}$$

The limit condition (95) therefore reduces to the form

$$\lim_{L\to\infty} \rho_c\left(\exp i \sum_{j=1}^k \xi_j \cdot \hat{\eta}_L(x_j)\right) = P\left(\exp i \sum_{j=1}^k \xi_j \cdot \eta^{(cl)}(x_j)\right),$$

or equivalently

$$\lim_{L\to\infty} \rho_c\left(\exp i \sum \xi(x) \cdot \hat{\eta}_L(x)\right) = P\left(\exp i \sum \xi(x) \cdot \eta^{(cl)}(x)\right),$$

i.e. by (88),

$$\lim_{L\to\infty} \mu_{c,L}(\xi) = \bar{\mu}(\xi), \tag{97}$$

† It is, nevertheless, conventionally referred to as the 'renormalization group', a term which is justified only in the case of continuous systems.

‡ To be precise, the function ϕ appearing in (96) should be absolutely summable (i.e. $\Sigma |\phi(\xi_1, \ldots, \xi_k)| < \infty$) in the first alternative and continuous and absolutely integrable (i.e. $\int |\phi(\xi_1, \ldots, \xi_k)| \, d\xi_1 \ldots d\xi_k < \infty$) in the second one.

where

$$\bar{\mu}(\xi) = P\left(\exp \mathrm{i} \sum_x \xi(x) \,.\, \eta^{(cl)}(x)\right), \tag{98}$$

the characteristic function for P. Moreover, by (93), the formula (97) is equivalent to

$$\lim_{L \to \infty} (R(L)\mu_{\mathrm{c}})(\xi) = \bar{\mu}(\xi),$$

which we shall write as

$$\lim_{L \to \infty} R(L)\mu_{\mathrm{c}} = \bar{\mu}. \tag{99}$$

This formula is therefore an expression of the central limit condition, (95), in terms of the action of the RS on the characteristic function μ_{c} governing the order field, $\hat{\eta}(x)$, in the state ρ_{c}. Furthermore, as we shall prove in Appendix B, it follows merely from (99) and the commutativity of the components of $\hat{\eta}_L(x)$ in the limit $L \to \infty$ that $\bar{\mu}$ is the characteristic function of a classical probability. In other words, the non-central limit condition (95) is equivalent to that of the convergence of $R(L)\mu_{\mathrm{c}}$ as $L \to \infty$.

It is now a simple consequence of the semigroup property (94) of the RS and the limit condition (99) that

$$R(L)\bar{\mu} = \bar{\mu}, \tag{100}$$

i.e. $\bar{\mu}$ is invariant under the RS or, to use a standard terminology, it is a *fixed point* of that semigroup. Moreover, by eqns (93), (98), and (99),

$$\bar{\mu} = \lim_{L \to \infty} \mu_{c,L} = \lim_{L' \to \infty} R(L')\mu_{c,L}. \tag{101}$$

This represents a *stability* property of $\bar{\mu}$, since it signifies that there are characteristic functions, $\mu_{c,L}$, that are 'close' to $\bar{\mu}$ for large L and are driven to that fixed point by the action of the RS. In other words, the formula (101) defines a sense in which $\bar{\mu}$ is a *stable fixed point* of the RS. Furthermore, since this semigroup corresponds to scale transformations, the fixed point property of $\bar{\mu}$ implies the *scale invariance* of the limit distribution P, its Fourier transform.

Thus, the assumption that the critical state ρ_{c} enjoys the non-central limit property (95) leads to a picture in which the large-scale properties of this state are represented by a scale invariant distribution, P, whose characteristic function, $\bar{\mu}$, is a stable fixed point of the RS. The theory of critical phenomena may therefore be centred on the study of fixed points of this semigroup according to the following programme.

I. Obtain the fixed points of the RS that satisfy the stability condition (101), for values of the normalization index in the range given by (86). Note that this does not involve any properties of the interactions, since $R(L)$ does not depend on these.

II. Select those stable fixed points, $\bar{\mu}$, given by scaling limits, $\lim_{L\to\infty} R(L)\mu$, of equilibrium characteristic functions of infinite systems, as in (99). This serves to identify the critical points of these systems and their limit distributions, P.

III. Investigate the action of the RS on equilibrium states close to critical points so as to obtain a microscopic version of Kadanoff's argument and to evaluate the indices, p and q, that appear there as phenomenological constants.

This is essentially a quantum mechanical version of Sinai's programme for implementing ideas expressed in a more heuristic form by Wilson (1971): in Sinai's (1982) classical treatment, the RS was expressed as a semigroup of transformations of probability distributions, rather than of the characteristic functions that are their Fourier transforms. We note here that, in view of the last observation of the above item (I), one can envisage that the RS treatment could provide a natural interpretation of the fact that critical properties of macroscopic systems depend only on general qualitative features of their microscopic constitutions. We shall now describe how the RS programme has been carried through by Bleher and Sinai (1973, 1975) for a tractable class of models, before returning to discuss results and problems of the general theory.

5.4.3. RS treatment of Dyson's hierarchical model

The model. The hierarchical model was originally constructed by Dyson (1969), who proved that it has a phase transition. Baker (1972) observed that it also enjoyed simple scaling properties, and these were subsequently exploited by Bleher and Sinai (1973, 1975) in their RS treatment of the model, which we shall now describe.

The model is a finite, but arbitrarily large system, Σ_N, consisting of 2^N Ising spins that occupy the segment $\Lambda_N \equiv \{n = 1, 2, \ldots, 2^N\}$ of a one-dimensional lattice with unit spacing. Λ_N is thus composed of Λ_{N-1} and its copy $\Lambda'_{N-1} = \{2^{N-1}+1, \ldots, 2^N\}$. Correspondingly, Σ_N consists of two coupled subsystems, namely Σ_{N-1} and its copy Σ'_{N-1}, which occupy the regions Λ_{N-1} and Λ'_{N-1}, respectively. It is assumed that the energy of interaction between these two halves of Σ_N is $-g^N\sigma_N^2$, where σ_N is the total spin of Σ_N and g is a positive constant. Hence, denoting the Hamiltonians and total spins of Σ_N, Σ_{N-1} and Σ'_{N-1} by (H_N, σ_N), (H_{N-1}, σ_{N-1}), and $(H'_{N-1}, \sigma'_{N-1})$, respectively,

$$H_N = H_{N-1} + H'_{N-1} - g^N\sigma_N^2 \tag{102}$$

where

$$\sigma_N = \sigma_{N-1} + \sigma'_{N-1} = \sum_{1}^{2^N} \sigma(n), \tag{103}$$

$\sigma(n)$ $(= \pm 1)$ being the spin at the site n. Since Σ'_{N-1} is assumed to be a copy of Σ_{N-1}, H'_{N-1} is simply the space translate of H_{N-1} over the distance 2^{N-1} and therefore eqn (102) expresses H_N in terms of H_{N-1}. We assume that formula to be valid for all values of N exceeding some finite number N_0. The formula (102) then serves to define H_N recursively in terms of H_{N_0}. Thus, the system Σ_N has two kinds of interactions, namely the short-range ones on which H_{N_0} and its copies are based, and the long-range ones represented by the last term in (102). It is these latter interactions that lead to the phase transition in Σ_N, in the limit $N \to \infty$. It will be assumed that H_{N_0}, and therefore also H_N, is invariant with respect to spin reversals.

It is necessary to impose restrictions on the value of g in order that the energy of interaction, $-g^N\sigma_N^2$, between Σ_{N-1} and Σ'_{N-1} acts as a 'surface effect', so that the thermodynamic potentials of Σ_N are well defined in the limit $N \to \infty$. Thus we assume that the norm of this interaction increases like $|\Lambda_N|^\alpha$ with N for some α between 0 and 1. Since $|\Lambda_N| = 2^N$ and $\|g^N\sigma_N^2\| = g^N 2^{2N}$, by (103), this is equivalent to assuming that

$$\tfrac{1}{4} < g < \tfrac{1}{2}. \tag{104}$$

The value of g represents essentially a power law governing the long-range interactions in the model. For it follows from (102), as applied for values of $N > N_0$, that the interaction energy between spins separated by a distance $r = 2^n$ is $\sim g^n = r^{-l}$, with $l = -\ln g/\ln 2$. The condition (104) therefore signifies that l lies between 1 and 2.

The canonical distribution function, representing the probability that the spin configuration of Σ_N at temperature T is $(\sigma(1), \ldots, \sigma(2^N))$, is

$$f_N(\sigma(1), \ldots, \sigma(2^N)) = Z_N^{-1} \exp(-H_N/kT), \tag{105}$$

where the partition function Z_N is given by the formula

$$Z_N = \sum_{\{\sigma(x)\}} \exp(-H_N/kT). \tag{106}$$

It follows immediately from eqns (102), (103), and (105) that f_N satisfies the following key recurrence relation

$$f_N(\sigma(1), \ldots, \sigma(2^N))$$
$$= \frac{Z_{N-1}^2}{Z_N} f_{N-1}(\sigma(1), \ldots, \sigma(2^{N-1})) f_{N-1}(\sigma(2^{N-1}+1), \ldots, \sigma(2^N)) \exp(g\sigma_N^2/kT). \tag{107}$$

In the presence of an external field y, the exponential in the formulae (105) and (106) becomes modified by a factor $\exp(y \,.\, \sigma_N/kT)$. Therefore the zero field susceptibility, defined as the derivative with respect to y, at $y = 0$, of the canonical expectation value of the spin density $\sigma_N/|\Lambda_N|$, is

$$\chi = \langle \sigma_N^2 \rangle_N / kT\, |\Lambda_N|, \tag{108}$$

where $\langle \ldots \rangle_N$ is the mean for the distribution f_N.

The macroscopic distribution. We describe the macroscopic properties of the model in terms of a single collective variable $\sigma_N/|\Lambda_N|^{m/2}$, which plays the same role as the block variables of the general theory. Since $|\Lambda_N| = 2^N$, this variable may be expressed as

$$\sigma_{N,\lambda} = \sigma_N/\lambda^{N/2} \tag{109}$$

with $\lambda = 2^m$, a constant. For non-central limiting behaviour of $\sigma_{N,\lambda}$ at the critical point, the bounds on λ, corresponding to those given by (86) for m in the general case, are therefore

$$2 < \lambda \leqslant 4. \tag{110}$$

We shall also have occasion to examine the statistics of $\sigma_{N,\lambda}$ for other values of λ, when we consider states slightly above or below the critical point. In particular, the value $\lambda = 2$ is appropriate for the analysis of the fluctuation variable $\sigma_N/\|\Lambda_N\|^{\frac{1}{2}}$ in situations where the central limit theorem is applicable, while $\lambda = 4$ is suitable for a treatment of states with a spontaneous magnetization corresponding to a non-zero expectation value of $\sigma_N/|\Lambda_N|$.

In order to formulate the equilibrium distribution function for $\sigma_{N,\lambda}$, we note that, by (103) and (109), the values taken by this variable are $\lambda^{-N/2} \times (-2^N, -2^N + 2, \ldots, 2^N)$. The probability that $\sigma_{N,\lambda}$ takes the value z, belonging to this set, when the system is in its canonical equilibrium state is then

$$P_N(z, \lambda) = \sum_{\sigma_{N\lambda} = z} f_N(\sigma(1), \ldots, \sigma(2^N)). \tag{111}$$

In fact, it follows immediately from this formula and (109) that the forms of $P_N(z, \lambda)$ for different values of λ are simply interrelated, as

$$P_N(z, \lambda) \equiv P_N(\lambda^{N/2} z, 1). \tag{112}$$

The canonical expectation values of functions of $\sigma_{N,\lambda}$ may be expressed in terms of the distribution $P_N(z, \lambda)$ by the formula

$$\langle F(\sigma_{N,\lambda}) \rangle_N = \sum^{(N)} P_N(z, \lambda) F(z), \tag{113}$$

the superscript (N) indicating that summation is taken over the eigenvalues of $\sigma_{N,\lambda}$.

The theory is centred on the properties of $P_N(z, \lambda)$ in the limit $N \to \infty$. Here we term a probability distribution $P(z, \lambda)$ the limit of $P_N(z, \lambda)$ if

$$\int P(z, \lambda) F(z) \, dz = \lim_{N \to \infty} \sum^{(N)} P_N(z, \lambda) F(z) \tag{114}$$

for bounded continuous $F(z)$. The objective of the theory is to establish the following results.

(a) The critical point is characterized by the convergence of $P_N(z, \lambda)$ to $P(z, \lambda)$, with λ satisfying the non-central limit condition (110).

(b) Above the critical point, the central limit theorem applies and the susceptibility diverges as $|t|^{-\beta}$, where the critical index β depends only on the long-range coupling constant g and not on the short-range interactions represented by H_{N_0}. In view of eqns (108) and (109), this central limit condition takes the form

$$P_N(z, 2) \sim \exp(-\text{const.}\ |t|^{\beta} z^2) \quad \text{for large } N, \text{ and } t > 0 \tag{115}$$

(c) Below the critical point, the spin density $\sigma_N/|\Lambda_N|$ is distributed according to the arithmetic mean of two probabilities centred at values $\pm m$ of the spontaneous polarization and carrying dispersions of the order of $\Lambda_N^{-\frac{1}{2}}$, so that

$$P_N(z, 4) \rightarrow \tfrac{1}{2}(\delta(z-m) + \delta(z+m)) \quad \text{as} \quad N \rightarrow \infty, \tag{116}$$

where

$$m \sim |t|^{\gamma} \tag{117}$$

and the critical index γ is independent of the short-range forces. We now proceed to the derivation of these results and the evaluation of the critical indices, β and γ.

The renormalization mapping. By eqns (107), (109), and (111), the probability $P_N(z, \lambda)$ satisfies the recurrence relation

$$P_N(z, \lambda) = L_N \exp((g\lambda)^N z^2/kT) \sum_{z'}^{(N-1)} P_{N-1}(z', \lambda) P_{N-1}(\lambda^{\frac{1}{2}} z - z', \lambda), \tag{118}$$

where L_N is a constant whose value is determined by the normalization of P_N and P_{N-1}. For non-central limiting behaviour, $P_N(z, \lambda)$ should tend to a distribution $P(z, \lambda)$ as $N \rightarrow \infty$, for a value of λ in the range given by (110). This restriction on λ implies that $\lambda > 2$ and therefore, as $g < \frac{1}{2}$, by (104), the only value of λ for which the exponential in (118) can converge, as $N \rightarrow \infty$, is g^{-1}. We therefore explore the limit distribution based on $\lambda = g^{-1}$, with the aim of subsequently confirming that this is the value of the normalization index λ that corresponds to the critical non-central limiting behaviour of P_N. Thus, putting $\lambda = g^{-1}$, we observe that the recurrence relation (118) reduces to the form

$$P_N(z, g^{-1}) = L_N \exp(z^2/kT) \sum_{z'}^{(N-1)} P_{N-1}(z', g^{-1}) P_{N-1}(g^{-\frac{1}{2}} z - z', g^{-1}). \tag{119}$$

We write this equation formally as

$$P_N = \mathcal{R} P_{N-1}. \tag{119$'$}$$

$\mathscr{R}$ is thus a non-linear mapping whose iterates form a semigroup $\{\mathscr{R}^n \mid n = 1, 2, \ldots,\}$ analogous to the RS of the general case. We term $\mathscr{R}$ the *renormalization mapping* and the semigroup that it generates the RS.

We now look for a limit $P(z, g^{-1})$ of the distribution $P_N(z, g^{-1})$. By (119), it follows that any such limit must satisfy the integral equation

$$P(z, g^{-1}) = L \exp(z^2/kT) \int_{-\infty}^{\infty} dz' P(z', g^{-1}) P(g^{-\frac{1}{2}}z - z', g^{-1}), \qquad (120)$$

where L is a normalization constant. As this equation is the limiting form of (119)′, we may write it formally as

$$P = \mathscr{R}P, \qquad (120)'$$

signifying that P is a *fixed point* of $\mathscr{R}$. Furthermore, as P is the limit of P_N, as $N \to \infty$, we see from (119)′ that

$$P = \lim_{N\to\infty} P_N \equiv \lim_{L\to\infty} \mathscr{R}^L P_N. \qquad (121)$$

This is a stability condition for P, since it signifies that the repeated action on distributions P_N, which are ‘close’ to P_N for large N, will drive them towards that limit. In this sense therefore, P is a *stable fixed point* of the RS.

Thus, as in the general theory, we see that the non-central limiting distribution, P, corresponds to a stable fixed point of the RS.

Determination of the stable fixed point. It is easy to check that the fixed point equation (120) ($\equiv$(120)′) has a Gaussian solution, namely

$$P_G(z) = \left(\frac{a}{\pi}\right)^{\frac{1}{2}} \exp(-az^2), \qquad (122)$$

with

$$a = ((1 - 2g)kT)^{-1} \quad \text{and} \quad L = 2^{\frac{1}{2}}. \qquad (123)$$

This is, of course, not a central limiting type of solution, even though it is Gaussian, since the normalization index $\lambda = g^{-1} > 2$. Also, there is no claim here that P_G is the only solution of (119), and it is, in fact, known that there are sometimes others, as we shall presently see. The important question, however, is whether P_G is the stable fixed point of the RS that corresponds to the limit of P_N at the critical point.

To answer this question, we study the stability of P_G by considering the action of $\mathscr{R}$ on neighbouring distributions of the form

$$P'(z, g^{-1}) = P_G(z) f(z), \qquad (124)$$

where

$$f(z) = 1 + \varepsilon\phi(z), \qquad (125)$$

and ε is small. Since $\mathscr{R}P'$ is given by the RHS of (120), with P' replacing P, it follows from (122)–(124) that

$$\mathscr{R}P' = \mathscr{P}_G \mathscr{Q} f, \tag{126}$$

where $\mathscr{Q}$ is the non-linear operator defined by

$$(\mathscr{Q}f)(z) = \left(\frac{2a}{\pi}\right)^{\frac{1}{2}} \int_{-\infty}^{\infty} du \, e^{-2au^2} f\left(\frac{z}{2g^{\frac{1}{2}}} + u\right) f\left(\frac{z}{2g^{\frac{1}{2}}} - u\right). \tag{127}$$

Hence, by (125),

$$\mathscr{Q}f = 1 + \varepsilon \mathscr{L}\phi + O(\varepsilon^2), \tag{128}$$

where

$$(\mathscr{L}\phi)(z) = \left(\frac{2a}{\pi}\right)^{\frac{1}{2}} \int_{-\infty}^{\infty} du \, e^{-2au^2} \left(\phi\left(\frac{z}{2g^{\frac{1}{2}}} + u\right) + \phi\left(\frac{z}{2g^{\frac{1}{2}}} - u\right)\right). \tag{129}$$

The stability of the fixed point P_G of the RS may therefore be studied in terms of this linear operator $\mathscr{L}$. In fact, it is known† that this is a self-adjoint operator, with discrete spectrum, in a certain Hilbert space,‡ and that its eigenfunctions and corresponding eigenvalues are $\{H_n(\alpha z)\}$ and $\{\gamma_n = 2(4g)^{-n/2}\}$, respectively, where the H_n's are the Hermite polynomials and α is a constant. Moreover, there is no need to look beyond the even values of n, since we are concerned here with the approach to a fixed point due to the iterated action of $\mathscr{R}$ on the distribution function P_N, which must be even, in view of the symmetry of Σ_N with respect to spin reversals. There is also no need to consider the eigenfunction $H_0(\alpha z)$, since this is merely a constant, whose contribution to ϕ in (125) would be absorbed into a normalization factor. Hence the relevant eigenfunctions of $\mathscr{L}$ are the even Hermite polynomials $\{H_{2n}(\alpha z)\}$ and the corresponding eigenvalues are

$$\gamma_{2n} = 2(4g)^{-n}, \qquad n = 1, 2, \ldots. \tag{130}$$

Now if, for example, $\phi(z) = H_{2n}(\alpha z)$, then by (126 and 128),

$$\mathscr{R}P = P_G(1 + \varepsilon\gamma_{2n} H_{2n}) + 0(\varepsilon^2),$$

and

$$\mathscr{R}^L P = P_G(1 + \varepsilon\gamma_{2n}^L H_{2n}) + 0(\varepsilon^2). \tag{131}$$

† See Bateman and Erdelyi (1955).
‡ This is the space of functions f such that

$$\int_{-\infty}^{\infty} |f(x)|^2 \exp\left(-\frac{(g-1)}{2g}\right) x^2 \text{ is finite.}$$

In this case, the repeated application of $\mathscr{R}$ leads back to the fixed point P_G if $|\gamma_{2n}|<1$ and leads away from it if $|\gamma_{2n}|>1$. Now as g lies between $\frac{1}{4}$ and $\frac{1}{2}$, by (104), it follows from (130) that γ_2 always exceeds unity, i.e. that H_2 represents an 'unstable direction'. It also follows that, if $g>\sqrt{2}/4$, then $|\gamma_4|$, $|\gamma_6|, \ldots$ are all less than unity and that therefore there are no other unstable directions. On the other hand, if $g<\sqrt{2}/4$, then $\gamma_4>1$. The stability properties are therefore different for the cases $g>\sqrt{2}/4$ and $g<\sqrt{2}/4$, and we therefore have to treat these two cases separately.

The case where $g>\sqrt{2}/4$. Here, as already noted, $\gamma_2>1$ and the other γ_{2n}'s all have magnitudes less than unity. To simplify the discussion, let us overlook, for the moment, the fact that P_N is a function of a discrete variable z, for finite N, and treat it formally as a continuous distribution function, i.e.

$$P_N = P_G\left(1+\sum_1^\infty \alpha_n H_{2n}\right), \tag{132}$$

where the α_n's depend on temperature and on N. Thus, in the linear approximation one finds that (cf. (131))

$$\mathscr{R}^L P_N = P_G\left(1+\sum_1^\infty \alpha_n \gamma_{2n}^L H_{2n}\right). \tag{133}$$

Since, in the present case, $\gamma_2>1$ and the remaining γ's have magnitudes less than unity, this formula implies that the iterated action of $\mathscr{R}$ on P_N leads to P_G *if and only if* $\alpha_1=0$. In view of the temperature dependence of α_1, one may anticipate that this latter condition serves to characterize the critical point. In other words, one may envisage that a rigorous analysis may reveal that $P_N(z,g^{-1})$ converges to $P_G(z)$ as $N\to\infty$ for precisely one value of the temperature, namely, the critical value. In fact, this has been proved to be the case by Bleher and Sinai (1973, 1975), who took full account of the discrete character of the probability P_N. Thus (cf. (122)),

$$P_N(z,g^{-1})\sim\exp(-az^2) \quad \text{for large } N\text{, at the critical point,}$$

or equivalently, by (112),

$$P_N(z,1)\sim\exp(-ag^N z^2) \quad \text{at the critical point.} \tag{134}$$

To pass from the critical point to the rest of the critical region, Bleher and Sinai (*loc. cit.*) devised an inductive technique to derive an asymptotic formula for $P_N(z,1)$ there, using a more precise version of (134) as a reference point. Their net result is that

$$P_N(z,1)\sim\exp(-(ag^N+a_1 2^{-N}t)z^2) \quad \text{for} \quad t>0 \text{ and large } N \tag{135}$$

and

$$P_N(z,1)\sim\exp(-(ag^N+a_1 2^{-N}|t|)(z-a_2|t|^{\frac{1}{2}})^2) + \exp(-(ag^N+a_1 2^{-N}|t|)(z+a_2|t|^{\frac{1}{2}})^2) \quad \text{for} \quad t<0 \text{ and large } N, \tag{136}$$

where a_1 and a_2 are positive constants. Since $g<\frac{1}{2}$, by (104), it follows from (112) that eqn (135) implies the following asymptotic form for $P_N(z,2)$ above the critical point:

$$P_N(z,2) \sim \exp(-a_1 t z^2) \quad \text{for} \quad t>0, \text{ for large } N, \tag{137}$$

which accords with the central limit theorem. On the other hand, as $g<\frac{1}{2}$, it follows from (112) and (136) that $P_N(z,4)$ has the following limiting behaviour below the critical point

$$P_N(z,4) \to \tfrac{1}{2}\delta(z-a_2|t|^{\frac{1}{2}}) + \tfrac{1}{2}\delta(z+a_2|t|^{\frac{1}{2}}) \text{ as } N\to\infty, \text{ for } t<0. \tag{138}$$

The last two equations serve to confirm the results anticipated in the formulae (115)–(117) and reveal that the critical indices β and γ are given by

$$\beta=\tfrac{1}{2}, \qquad \gamma=\tfrac{1}{2}. \tag{139}$$

These are, in fact, the values given by *mean field theories* of the Landau–Van der Waals type. Thus we conclude that, for $g>\sqrt{2}/4$, the critical indices take the mean field theoretical values, *irrespective of the details of the short-range interactions*.

The case where $g<\sqrt{2}/4$. This case is different from the previous one in the important respect that the operator $\mathscr{L}$ has at least one eigenvalue, γ_4, other than γ_2, that exceeds unity. Consequently, the H_4 direction becomes unstable and so the Gaussian distribution P_G no longer has the stability required of a limit of $P_N(z,g^{-1})$. A second fixed point was obtained by Bleher and Sinai (*loc.cit.*) by a method based on an idea due to Wilson and Fisher (1972). The method is to seek a non-Gaussian solution of the fixed point equation (120) in the form of a power series

$$P_G^{(\varepsilon)}(z) = P_G(z)\Big(1+\sum_1^{\infty} \varepsilon^n f_n(z)\Big) \tag{140}$$

where

$$\varepsilon = \frac{2}{4} - g \tag{141}$$

is the expansion parameter, which is positive in the present case. Bleher and Sinai were able to obtain such a solution by a perturbative method, and to show that it possesses the stability properties required of a limit of P_N at the critical temperature. In other words, the stable fixed point of $\mathscr{R}$ changes from the Gaussian one, P_G, to the non-Gaussian one, $P_G^{(\varepsilon)}$, as ε changes from negative to positive values. This situation is analogous to that frequently encountered in the theory of dynamical systems,† where the form of stable equilibrium states may change when some control parameter passes through a critical value.

† A simple example, cited by Cassandro and Jona-Lasinio (1978), is that of a dynamical system represented by the differential equation $dx/dt=-x(\varepsilon+x)$, whose stable equilibrium position (fixed point!) is $x=0$ when $\varepsilon>0$ and $x=-\varepsilon$ when $\varepsilon<0$.

Having obtained the non-Gaussian fixed points $P_G^{(\varepsilon)}$, Bleher and Sinai again employed their inductive technique to obtain asymptotic formulae for P_N in the critical region. In this way, they then obtained the critical indices β and γ as power series in ε, that did not depend on the short-range interactions of the model.

Thus, the Bleher–Sinai treatment of the hierarchical model, based on the stable fixed points of the RS, leads to a picture in which the critical indices reduce to those of the Landau–Van der Waals mean field theory when ε $(\equiv \sqrt{2}/4 - g) < 0$ and deviate from these values by power series in ε when this parameter is positive. Further, in all cases, these indices do not depend on the short range interactions of the model.

5.4.4. Discussion of the general RS theory

The programme for the general RS theory, set out at the end of §5.4.2, is of course immeasurably more complex than anything involved in the solution of the hierarchical model. At a rigorous level, Sinai (1982, Ch. 3) has constructed explicit Gaussian and non-Gaussian limit distributions corresponding to stable fixed points of the RS for block variables of classical systems by a procedure analogous to the one we have just described for the hierarchical model. This construction is also applicable to quantum systems, as their limit distributions are classical, as we noted in §5.4.4. However, although it has the merit of exhibiting the large-scale structure of critical states, Sinai's general construction provides no explicit correspondence between the limit distribution and the microscopically defined interactions of a system. Instead, it relates the limit distribution to effective interactions of the large scale block variables in the critical state. There is therefore a serious gap, in the general rigorous theory, between the microscopic specifications of the forces and the large-scale description of critical states in terms of stable fixed points of the RS.

At the more heuristic level, however, the correspondence between the microscopic and phenomenological descriptions is generally formulated by expressing the RS as transformations of the interactions rather than the states, by a procedure that may be schematically described as follows. Suppose that Σ is an arbitrarily large, but finite, spin system, say, whose interactions are represented by a set of coupling parameters J: in general the interactions may include some of the many-body type. The Hamiltonian will then be a function, $H(J)$, of the interaction parameters, and so too will be the canonical density matrix, $\exp(-\beta H(J))/\mathrm{Tr(idem)}$. If one now divides the system into cells of side L, as previously, and defines $\sigma_L(x)$ as the total spin in the cell centred at Lx, then the equilibrium properties of these block spins may be defined in terms of an effective Hamiltonian, $H(J_L)$, that is a function of the σ_L's and a new set of interaction parameters, J_L, defined by the condition that the canonical density matrix $\exp \times (-\beta H(J_L))/\mathrm{Tr(idem)}$ yields the same expectation values as the original one,

$\exp(-\beta H(J))/\text{Tr(idem)}$, for the observables generated by the σ_L's. In this way, one obtains a transformation $J \rightarrow J_L \equiv \mathcal{R}_L J$ of the interactions, the form of $\mathcal{R}_L$ being taken in the limit where the volume of the system becomes infinite. The transformations $\{\mathcal{R}_L\}$ then form a semigroup, which we shall refer to as the RS.

A vast body of research,† based on this or similar versions of the RS, has been remarkably successful, both in computing critical indices and in classifying macroscopic systems according to their critical properties. Here, it should be emphasized, the treatments are based on approximative schemes for obtaining explicit formulae for $\mathcal{R}_L$. As in the Bleher–Sinai analysis of the hierarchical model, the method employed is centred on the properties of the stable fixed points of the RS. Furthermore, for d-dimensional systems with short range interactions, the parameter $\varepsilon = (4 - d)$ plays the same role as the one denoted by ε in the Bleher–Sinai theory, described above: in both cases, the critical indices take the values given by the Landau–Van der Waals theory for $\varepsilon < 0$ and deviate from these values by power series expansions in ε for $\varepsilon > 0$.

Thus, the renormalization theory has provided a 'new' way of looking at critical phenomena and, at a non-rigorous level, a powerful technology for relating these to microscopic properties of systems. However, as pointed out by Griffiths (1981), there are still some serious unresolved mathematical and conceptual problems connected with the RS theory. Griffith's argument is that the renormalization, $\mathcal{R}_L$, of the interactions may contain singularities that are completely missed by the approximative techniques that are used to formulate this transformation. For example, if Σ is a system of Ising spins, each configuration of block spins would represent some space-dependent 'inner field'. The Helmholtz potential of Σ would then be a function of this field which might possess singularities, corresponding to phase transitions, in which case the renormalization $\mathcal{R}_L$ could also be singular. Griffiths has in fact provided examples where such singularities are shown to occur. The significance of this is that the methods used to analyse the stability of the fixed points of the RS are based on the assumption that $\mathcal{R}_L$ is appropriately smooth, or non-singular. There is therefore a problem of whether a rigorous treatment would reveal that the properties of the stable fixed points are unaffected by whatever singularities $\mathcal{R}_L$ may possess. Until that problem is solved one cannot say for sure whether the RS theory properly represents the microscopic mechanisms governing critical phenomena.

5.5. Concluding remarks

The developments discussed in this chapter provide a broad, though very incomplete, picture of the factors governing the occurrence of order–disorder transitions and the forms of critical singularities.

† See, for example, the review articles by Wilson and Kogut (1974) and Fisher (1974), and the paper by Niemeyer and Van Leeuwen (1974).

Let us now briefly discuss some of the outstanding problems that remain in the theory of phase transitions. First, we remark that the techniques that have been devised to prove the existence of transitions in certain classes of models suffer from two serious limitations, namely that (cf. §5.3) (a) they are confined to lattice systems and (b) they are largely dependent on reflection positivity arguments, that are inapplicable to some systems of physical interest, such as the Heisenberg ferromagnet. Evidently it is desirable to extend the present methodology so as to remove both of these limitations. This could be very difficult because of the problems in gaining mathematical control over the relevant correlation inequalities in situations where the local energy observables may be unbounded or where the reflection symmetry giving rise to RP is lacking.

As regards the theory of critical singularities, it is clear that the renormalization methods have provided an effective way of relating them to the large-scale properties of macroscopic systems. From a fundamental standpoint, however, the RS theory of these singularities must be regarded as incomplete at present, because the general connection between the interactions of a system and the large-scale limit of its critical distribution function has not been fully established, either by the rigorous probabilistic version of the theory or by the more heuristic and practical Hamiltonian version of it (cf. §5.4.4).

Finally, to put the developments in the theory of phase transitions into proper perspective, we should mention that the present statistical mechanical methodology provides no explanation of the empirical fact that practically every simple substance has solid, liquid, and gaseous phases. Indeed, one of the outstanding problems of the theory is the very existence of the solid phase, which presumably corresponds to a breakdown of Euclidean translational symmetry in favour of some crystallographic structure.

Appendix A
Proof that Griffiths's condition (3) implies symmetry breakdown

For simplicity, we shall confine the proof to the case where Σ is a system of Ising spins, that is symmetric with respect to spin reversals, and $\hat{\eta}(x)$ is the spin, $\sigma(x)$, at the site x. The general proof, for systems where $\sum_{x\in\Lambda}\hat{\eta}(x)$ commutes with the local Hamiltonian $H(\Lambda)$, follows analogously.

Thus, for an infinite Ising system, Σ, with spin reversal symmetry, we denote by $\phi(h, T)$ the Gibbs potential, as a function of the external magnetic field, h, and the temperature, T. The spontaneous magnetization at that temperature is then

$$m_0 = -\lim_{h\to+0}\frac{\partial\phi(h,T)}{\partial h}. \tag{A.1}$$

Griffith's theorem, which we shall now prove, is that

$$m_0^2 \geq \lim_{\Lambda\uparrow} |\Lambda|^{-2} \langle \sigma(\Lambda)^2 \rangle_\Lambda, \tag{A.2}$$

where

$$\sigma(\Lambda) = \sum_{x\in\Lambda} \sigma(x) \tag{A.3}$$

and $\langle \ldots \rangle_\Lambda$ denotes the equilibrium expectation value for the finite version $\Sigma(\Lambda)$, of Σ, in the absence of any external field. This theorem therefore implies that *if* condition (3) is satisfied, i.e. if

$$\lim_{\Lambda\uparrow} |\Lambda|^{-2} \langle \sigma(\Lambda)^2 \rangle_\Lambda > 0,$$

then $m_0 \neq 0$ and that, consequently, there is a breakdown of spin reversal symmetry. The proof of (A.2) is based on the following two lemmas.

***Lemma* 1.** Given $m^* > m_0$, one can find a polarization value m' and an external field h', such that

$$m_0 \leq m' < m^* \tag{A.4}$$

and

$$\phi(h', T) \geq \phi(0, T) - m'h' \tag{A.5}$$

***Lemma* 2.** Given $m^* > m_0$, the equilibrium probability $\pi_\Lambda(m^*)$, that the polarization of $\Sigma(\Lambda)$ is greater than or equal to m^* in the absence of an external field, is majorized by $\exp(-\beta\delta\,|\Lambda|)$ for some positive δ, which depends on m^* but not on Λ.

***Proof of Lemma* 1.** Since $\phi(h, T)$ is concave in h, it follows that there is a sequence $\{h_n\}$, tending to zero from above, such that $m_n \equiv -(\partial\phi/\partial h)(h_n, T) \to m_0$ as $n \to \infty$. Hence, if $m^* > m_0$, the value of m_n will be less than m^* for sufficiently large n. Furthermore, the concavity of ϕ ensures that

$$\phi(0, T) - \phi(h_n, T) \leq -h_n \frac{\partial \phi}{\partial h}(h_n, T) \equiv m_n h_n.$$

Hence, choosing $h' = h_n$ and $m' = m_n$, we have the required result.

***Proof of Lemma* 2.** Let $p_\Lambda(m, h)$ be the equilibrium probability that the polarization of $\Sigma(\Lambda)$, in the presence of an external field h, takes the value m, i.e.

$$p_\Lambda(m, h) = \sum_{\sigma(\Lambda)=m|\Lambda|} \exp -\beta(H(\Lambda) - h\sigma(\Lambda))/\exp(-\beta\Phi_\Lambda(h, T)), \tag{A.6}$$

where $\Phi_\Lambda(h, T)$ is the Gibbs free energy of $\Sigma(\Lambda)$. Therefore,

$$p_\Lambda(m, h) = p_\Lambda(m, 0) \exp \beta(\Phi_\Lambda(h, T) - \Phi_\Lambda(0, T) + mh\,|\Lambda|). \tag{A.7}$$

On the other hand, the total probability that the polarization does not fall below m^*, in the absence of any external field, is

$$\pi_\Lambda(m^*) = \sum_{m \geqslant m^*} p_\Lambda(m, 0). \tag{A.8}$$

On summing the formula (A.7) over values of $m \geqslant m^*$, we now see that the LHS of the resultant equation cannot exceed unity while, for $h > 0$, the RHS cannot fall below $\pi_\Lambda(m^*) \exp \beta(\Phi_\Lambda(h, T) - \Phi_\Lambda(0, T) + m^*h\,|\Lambda|)$. Hence

$$\pi_\Lambda(m^*) \leqslant \exp - \beta(\Phi_\Lambda(h, T) - \Phi_\Lambda(0, T) + m^*h\,|\Lambda|), \quad \text{for } h > 0. \tag{A.9}$$

Furthermore, by Lemma 1,

$$\phi(h', T) - \phi(0, T) + m'h' \geqslant 0 \tag{A.10}$$

for some $m_0 \leqslant m' < m^*$ and $h^* > 0$. Hence, as $\phi(h, T)$ is the infinite volume limit of $\Phi_\Lambda(h, T)/|\Lambda|$, it follows that, given $\delta > 0$,

$$|\Lambda|^{-1}(\Phi_\Lambda(h', T) - \Phi_\Lambda(0, T)) + m'h' > -\delta \tag{A.11}$$

for sufficiently large Λ. Choosing $\delta = \frac{1}{2}(m^* - m)h'$, it follows from this inequality and (A.9) that $\pi_\Lambda(m^*) \leqslant \exp(-\beta\delta\,|\Lambda|)$, as required.

Proof of the inequality (*A.2*). Since $p_\Lambda(m, 0)$ is the equilibrium probability that the polarization of $\Sigma(\Lambda)$ is m, in the absence of any external field, the RHS of (A.2) may be expressed in the form

$$\begin{aligned}|\Lambda|^{-2}\langle\sigma(\Lambda)^2\rangle_\Lambda &= \sum_m m^2 p_\Lambda(m, 0)\\ &= \sum_{m<m^*} m^2 p_\Lambda(m, 0) + \sum_{m \geqslant m^*} m^2 p_\Lambda(m, 0)\\ &\leqslant m^{*2} \sum_{m<m^*} p_\Lambda(m, 0) + \sum_{m \geqslant m^*} m^2 p_\Lambda(m, 0)\end{aligned}$$

Here, the first sum on the RHS cannot exceed unity, and, as the maximum value of m^2 is unity, the second sum cannot exceed $\pi_\Lambda(m^*)$ as defined by (A.8). Hence,

$$\begin{aligned}|\Lambda|^{-2}\langle\sigma(\Lambda)^2\rangle_\Lambda &\leqslant m^{*2} + \pi_\Lambda(m^*)\\ &\leqslant m^{*2} + \mathrm{e}^{-\beta|\Lambda|\delta}, \text{ by Lemma 2.}\end{aligned}$$

Therefore, $\lim_{\Lambda\uparrow} |\Lambda|^{-2}\langle\sigma(\Lambda)^2\rangle_\Lambda \leqslant m^{*2}$ for any $m^* > m_0$, which implies the required result (A.2).

Appendix B
Proof that $\lim_{L\to\infty} R(L)\mu_c$ is a classical characteristic function

In order to prove that $\bar{\mu}$ $(\equiv \lim_{L\to\infty} R(L)\mu_c)$ is classical, we invoke Bochner's theorem.† This tells us that the conditions for $\bar{\mu}(\xi)$ to be classical

† See Lukacs (1970), p. 71.

are that

($\bar{\mu}$.1) $\bar{\mu}(0) = 1$,

($\bar{\mu}$.2) $\bar{\mu}(\xi)$ is continuous w.r.t. each $\xi(x)$, and

($\bar{\mu}$.3) for any set $\xi^{(1)}, \ldots, \xi^{(l)}$ of ξ-fields and complex numbers $c_1, \ldots, c_l$,

$$\sum_{\alpha,\beta=1}^{l} \bar{c}_\alpha c_\beta \, \bar{\mu}(\xi^{(\alpha)} - \xi^{(\beta)}) \geqslant 0. \tag{B.1}$$

In fact, it follows immediately from (88) and (97) that ($\bar{\mu}$.1) is satisfied. The proof that $\bar{\mu}$ satisfies the conditions ($\bar{\mu}$.2,3) will be based on the following two properties of the block variables, $\hat{\eta}_L(x)$, which constituted the assumption of non-central limiting behaviour of the critical state ρ, in §5.4.

(a) The dispersion in $\hat{\eta}_L(x)$ for the state ρ_c tends to a finite, non-zero limit, K, as $L \to \infty$, i.e.

$$\lim_{L\to\infty} \rho_c((\hat{\eta}_L(x))^L) = K \tag{B.2}$$

(b) The components, $(\hat{\eta}_{L,1}(x), \ldots, \hat{\eta}_{L,n}(x))$ of $\hat{\eta}_L(x)$ commute in the limit $L \to \infty$, i.e.

$$\lim_{L\to\infty} \|[\hat{\eta}_{L,r}(x), \hat{\eta}_{L,s}(x)]\| = 0. \tag{B.3}$$

We shall also utilize the following operator-theoretic lemma, whose proof we shall defer until the end of this Appendix.

Lemma. *For any bounded local observables A, and B,*

$$\|[e^{i(A+B)} - e^{iA}e^{iB}]\| \leqslant \tfrac{1}{2}\|[A, B]\| \tag{B.4}$$

***Proof of* ($\bar{\mu}$.2).** By eqns (87)′, (97), (B.3), and (B.4), we may re-express $\bar{\mu}$ in the form

$$\bar{\mu}(\xi) = \lim_{L\to\infty} \mu'_{c,L}(\xi), \tag{B.5}$$

where

$$\mu'_{c,L}(\xi) = \rho_c(V_L(\xi) \exp i(\xi_r(x) \,.\, \hat{\eta}_{L,r}(x))), \tag{B.6}$$

and

$$V_L(\xi) = \exp i\left(\sum_{x' \neq x} \xi(x') \,.\, \eta_L(x') + \sum_{s \neq r} \xi_s(x) \,.\, \eta_{L,s}(x)\right). \tag{B.7}$$

Hence, by (B.6),

$$\frac{\partial \mu'_{c,L}}{\partial \xi_r(x)} = i\rho_c(V_L(\xi) \exp i(\xi_r(x) \,.\, \hat{\eta}_{L,r}(x))\hat{\eta}_{L,r}(x))$$

and therefore, by the Schwartz inequality,

$$\left|\frac{\partial \mu'_{c,L}}{\partial \xi_r(x)}\right|^2 \leqslant \rho_c((\hat{\eta}_{L,r}(x))^2) \leqslant \rho_c((\hat{\eta}_L(x))^2).$$

From this inequality and (B.2), it follows that the derivatives $\partial\mu'_{c,L}/\partial\xi_r(x)$ are all uniformly bounded. Consequently, the limit $\bar{\mu}(\xi)$ of $\mu'_{c,L}(\xi)$ is continuous in all its arguments, as required.

***Proof of* ($\bar{\mu}$.3).** Since ρ_c is a state and hence a *positive* functional of the observables,

$$\rho_c\left[\left(\sum_{\alpha=1}^{l} c_\alpha \exp -\mathrm{i}\left(\sum \xi^{(\alpha)}(x)\,.\,\hat{\eta}_L(x)\right)\right)^* \times\left(\sum_{\alpha=1}^{l} c_\alpha \exp -\mathrm{i}\left(\sum \xi^{(\alpha)}(x)\,.\,\hat{\eta}_L(x)\right)\right)\right]\geqslant 0.$$

Hence,

$$\sum_{\alpha,\beta=1}^{l} \bar{c}_\alpha c_\beta \rho_c\left(\exp \mathrm{i}\left(\sum \xi^{(\alpha)}(x)\,.\,\hat{\eta}_L(x)\right)\exp -\mathrm{i}\left(\sum \xi^{(\beta)}(x)\,.\,\hat{\eta}_L(x)\right)\right)\geqslant 0.$$

By (B.3) and (B.4), this inequality reduces to the following form in the limit $L\to\infty$.

$$\lim_{L\to\infty}\sum_{\alpha,\beta=1}^{l} \bar{c}_\alpha c_\beta \rho_c\left(\exp \mathrm{i}\left(\sum\left(\xi^{(\alpha)}(x)-\xi^{(\beta)}(x)\right)\,.\,\hat{\eta}_L(x)\right)\right)\geqslant 0,$$

i.e. by (87)′ and (93),

$$\sum_{\alpha,\beta=1}^{l} \bar{c}_\alpha c_\beta \bar{\mu}(\xi^{(\alpha)}-\xi^{(\beta)})\geqslant 0,$$

which is the required result.

***Proof of Lemma*.** Since the derivatives of $e^{i(A+B)t}$, e^{-iAt} and e^{-iBt} with respect to t are $i\,e^{i(A+B)t}(A+B)$, $-iAe^{-iAt}$ and $-iBe^{-iBt}$, respectively, it follows that

$$\frac{\mathrm{d}}{\mathrm{d}t}(e^{i(A+B)t}e^{-iBt}e^{-iAt})\equiv ie^{i(A+B)t}e^{-iBt}(e^{iBt}Ae^{-iBt}-A)e^{-iAt}$$

$$\therefore\ e^{i(A+B)}e^{-iB}e^{-iA}-I=i\int_0^1 \mathrm{d}t\ e^{i(A+B)t}e^{-iBt}(e^{iBt}Ae^{-iBt}-A)e^{-iAt}$$

$$=\int_0^1 \mathrm{d}t\ e^{i(A+B)t}e^{-iBt}\left(\int_0^t \mathrm{d}s\ e^{iBs}[B,A]e^{-iBs}\right)e^{-iAt}.$$

Hence

$$\|e^{i(A+B)}e^{-iB}e^{-iA}-I\|\leqslant\int_0^1 \mathrm{d}t\int_0^t \mathrm{d}s\,\|[B,A]\|=\tfrac{1}{2}\|[A,B]\|,$$

or equivalently,

$$\|e^{i(A+B)}-e^{iA}e^{iB}\|\leqslant\tfrac{1}{2}\,\|[A,B]\|,$$

as required.

6

Metastable states

6.1. Introduction

Metastable states of macroscopic systems are ones that behave as equilibrium states, except for the fact that they do not correspond to absolute minima of free energy. Their characteristic properties, as seen from the empirical standpoint, are the following.

(M1) *They do not minimize the free energy.*

(M2) *They nevertheless have 'good' thermodynamical properties, satisfying both the Zeroth Law and the restricted version of the Second Law given by the standard relationship between entropy, S, internal energy, U, and mechanical work, W, namely*

$$T\,\mathrm{d}S = \mathrm{d}U + \mathrm{d}W. \tag{1}$$

(M3) *They have 'very long' lifetimes, even when coupled to thermal reservoirs.*

Furthermore, *some* metastable states, e.g. those corresponding to superheated or supercooled phases of a system, possess the following additional property.

(M4) *The thermodynamic functions of a metastable state are (or closely approximate to) smooth continuations, beyond a phase transition, of those of some equilibrium phase.*

There is an abundance of metastable states in nature: arguably, they are more prevalent than true equilibrium states. Familiar examples are the superheated or supercooled phases of matter, current-carrying states of superfluids, states of ferromagnets with polarizations opposed to sufficiently weak magnetic fields, and allotropes of some materials whose free energies are not minimal – here, diamond (!) is a dramatic example. Furthermore, non-spherical solid objects are generally metastable as their surface areas, and hence their surface energies, are higher than those of their spherical counterparts. This means that virtually all one's utensils, e.g. tables, chairs, books, etc., are metastable. For similar reasons, so too are states of systems that are separated into different phases in contiguous regions, e.g. as in a ferromagnet with domain structure: for such states are metastable because of the surface energy between the different phases. Finally, it is worth remarking that, from the standpoint of Nuclear Physics, most of the materials normally classed as being in equilibrium are really metastable. One sees this by looking, for example, at the materials containing equal numbers of protons, neutrons, and electrons, e.g. helium, carbon, nitrogen, etc. For since these materials have quite different thermal properties, one of them generally has lower free energy than the rest, which therefore correspond to metastable states of systems of protons, electrons, and neutrons.

The basic statistical mechanical problem of metastability is that of the existence of non-equilibrium states with good thermodynamical properties and very long lifetimes. Penrose and Lebowitz (1971) have made the important observation that, for any system, Σ, this problem may be reduced to that of finding a suitable subset, $\Omega^{(0)}$ of states, containing no absolute minima of the free energy, such that the dynamics of Σ permits only a very slow rate of escape from $\Omega^{(0)}$. For then, under mild supplementary conditions on $\Omega^{(0)}$, the state, $\rho^{(0)}$, that minimizes the free energy over this reduced state space will have good thermodynamical properties for the same reasons that the absolute minimization of free energy leads to standard thermodynamics, while the slow escape rate from $\Omega^{(0)}$ ensures that $\rho^{(0)}$ has a long lifetime. A different observation, due to the present author (1976, 1980) is that locally stable states, as defined in Chapter 3, have *infinite* lifetimes,† even in the face of coupling to thermal reservoirs. Furthermore (cf. Chapter 3), states that are locally, but not globally, stable can exist in certain systems with long-range forces.

In view of these considerations, it is clear that the metastability conditions (M1–3) will be satisfied by states that are locally, but not globally, stable and that minimize the free energy over some reduced state space. We term such states *ideally metastable*. Since they have infinite lifetimes, due to their local stability, they satisfy the conditions (M1–3) in the most decisive way possible. However, translationally invariant states of this kind can occur only in systems with forces of sufficiently long range for the interaction energy between the particles inside and outside a 'large' bounded region to be a volume effect, rather than a surface effect: for otherwise the local and global stability conditions become equivalent.‡ On the other hand, non-translationally invariant ideal metastable states can occur even in systems with short-range forces, as we shall presently see (in §6.2.2). These are states with different phases occupying contiguous spatial regions, as in a ferromagnet with a domain structure. Their excess free energies, above equilibrium values, correspond to surface, rather than volume, effects.

Since ideal metastability occurs in only rather limited circumstances, it is clear that most metastable states must be non-ideal and so must lack the

† See comment at the end of §3.7.1.
‡ See Chapter 3, Appendix D.

local stability property: instead they must have sufficient stability to ensure 'very long', rather than infinite, lifetimes. Thus, we define a non-ideal, or *normal, metastable state* to be a state that minimizes the free energy over a reduced state space, $\Omega^{(0)}$, and whose lifetime though finite, is very long, in some appropriate sense. The problem of determining the factors governing such lifetimes is evidently a dynamical one.

The object of the present chapter is to provide a framework for the theory of metastability, based on the above specifications of ideal and normal metastable states. In §6.2, we shall set up the theory of ideal metastability and provide some tractable model examples of it. In §6.3, we shall formulate the theory of normal metastability, based largely on the idea that the relevant stabilization occurs as a result of energy barriers, created by the collective action of the microscopic components of the system, that obstruct the escape from the reduced state space $\Omega^{(0)}$. As an example, we shall treat the metastability of states of an Ising ferromagnet† polarized in opposition to a weak external field: there the barriers against spin reversals are due to the 'inner polarization field'. In §6.4, we shall briefly summarize the picture that emerges from the theory described here.

6.2. Ideal metastability

We define a state, $\rho^{(0)}$, of an infinite system, Σ, to be ideally metastable if

(IM1) *it is locally stable*,

(IM2) *it is not globally stable*, *and*

(IM3) *it minimizes the free energy over a reduced set*, $\Omega^{(0)}$, *of states*, *that is closed with respect to space translations and contains the limits of its sequences of translationally invariant states*.

In the case of states with only limited translational symmetry, e.g. over two spatial directions instead of three, the specifications of $\Omega^{(0)}$ in (IM3) are modified in a straightforward way.

It follows from (IM3) that ideal metastable states enjoy thermodynamical properties, represented by the standard relation (1), for precisely the same reasons as the globally stable states do. Furthermore, the local stability property (IM1), which is equivalent to that of KMS, ensures that ideal metastable states have infinite lifetimes, even when coupled to thermal reservoirs. This may be interpreted as signifying that the counterparts of these states for finite systems have lifetimes that increase indefinitely with volume.

As indicated in the previous Section, the translationally invariant ideal metastable states are different from the non-translationally invariant ones in

† The theory of these states is due originally to Capocaccia, Cassandro, and Olivieri (1974).

some important respects, and so we shall treat these two classes of states separately.

6.2.1. Translationally invariant states

Ideally metastable translationally invariant states must evidently be limited to the systems with interactions of sufficiently long range to support locally stable, translationally invariant states that are not also globally stable. We shall now provide two examples of such states in rather simple models, before briefly discussing their possible existence in gravitational systems, such as Neutron Stars.

Example 1. *Cluster model with many-body forces*. This is the model of §3.6.1, which, as we showed there, supported translationally invariant states that were locally but not globally stable. Let us briefly recall the principal features of this model.

It is a system of atoms on a lattice† X, with many-body forces that favour condensation into a reduced state space, $\Omega^{(0)}$, based on a restricted set of atomic states. Specifically, these forces are such that an infinite energy is required to remove the system from the state subspace $\Omega^{(0)}$ by localized modifications of state. As a consequence of this property of the forces, the state, $\rho^{(0)}$, that minimizes the free energy density over $\Omega^{(0)}$ is locally stable. However, it is not globally stable at sufficiently high temperatures, because of its entropy deficiency. Therefore, at such temperatures, $\rho^{(0)}$ satisfies the conditions (IM1–3) and so is ideally metastable.

Note. We have also proved elsewhere (Sewell 1976) that, in the case of Fisher's (1967*a*) classical cluster model, the analogue of this state $\rho^{(0)}$ corresponds to a superheated liquid phase, whose Gibbs potential is an analytic continuation of that of a low temperature equilibrium phase.

Example 2. *Mean field theoretic models*. These are models with weak, long-range, two-body forces that lead, in an appropriate limit, to phase transitions governed by self-consistent mean field theories of the Van der Waals and Weiss type. For definiteness, we shall focus our attention here on a prototype example; namely, the Ising–Weiss chain. This is a system, Σ_N, of N Ising spins in consecutive sites of a one-dimensional lattice, with interaction energy $-(J/N)\sigma(n)\sigma(n')$ between the spins at arbitrary sites n and n', J being positive. Thus, the Hamiltonian for the model in the presence of a uniform external field h is

$$H_N = -\frac{J}{2N}\sum_{n,n'=1}^{N}\sigma(n)\sigma(n') - h\sum_{n=1}^{N}\sigma(n), \tag{2}$$

and each spin experiences the action of the external field h and an internal

† Although we formulated the lattice as one-dimensional in Chapter 3, there is no difficulty in generalizing the theory to arbitrary dimensionality (cf. Sewell 1977*b*).

one Jm_N, where m_N is the polarization observable

$$m_N = N^{-1} \sum_{1}^{N} \sigma(n). \tag{3}$$

One can therefore re-express H_N in terms of this latter observable by the formula

$$H_N = -\sum_{n=1}^{N} (h + Jm_N)\sigma(n) + \tfrac{1}{2}JNm_N^2. \tag{4}$$

Emch (1967*a*) has demonstrated that the properties of the model simplify drastically in the limit $N \to \infty$. The essential reason is that m_N then reduces to a *c*-number, m, in any state with the pure phase property of dispersionless global intensive variables. Hence, by (4), Σ_N reduces to a system of independent spins in the field $(h + Jm)$, with self-energy density $\frac{1}{2}Jm^2$, in the limit $N \to \infty$. Consequently, as the free energy density of a single Ising spin in a field h' is $-kT \ln(2 \cosh \beta h')$, it follows that the equilibrium free energy density of Σ is

$$\phi(h, T) = \min \phi_m(h, T), \tag{5}$$

where

$$\phi_m(h, T) = -kT \ln(2 \cosh \beta(h + Jm)) + \tfrac{1}{2}Jm^2. \tag{6}$$

The extrema of ϕ_m, considered as a function of m, for fixed h and T, occur when

$$m = \tanh \beta(h + Jm), \tag{7}$$

which is the standard Weiss self-consistency equation for m. For $T > T_c \equiv J/k$, this equation has just one solution and that corresponds to a minimum for ϕ_m. On the other hand, for $T < T_c$, and $|h|$ sufficiently small, it has three solutions, corresponding to two minima, one positive and one negative, separated by a maximum (cf. Fig. 8). Furthermore, if $h \neq 0$, the minimum, m, with the same sign as h is absolute, and the other one, with opposite sign to h, strictly local.

We shall now show that this latter minimum corresponds to an ideal metastable state – evidently one whose polarization opposes the external field. For this we first recall that, as noted above, the model reduces to a system of independent spins in a magnetic field $(h + Jm)$, with self-energy density $\frac{1}{2}Jm^2$. Hence, for *fixed* m, the state ρ_m that minimizes the free energy density is simply the equilibrium state of a system of independent spins in the field $h + Jm$. Suppose now that $m^{(0)}$ $(= m^{(0)}(h, T))$ is the value of m corresponding to the local non-absolute minimum of ϕ_m. Then $\phi_{m^{(0)}}$ is the equilibrium state of the system $\Sigma^{(0)}$ of independent spins in the field $h + Jm^{(0)}$. It is therefore also a locally stable one of the model Σ, since the effective field is this system is determined by the global space average of the

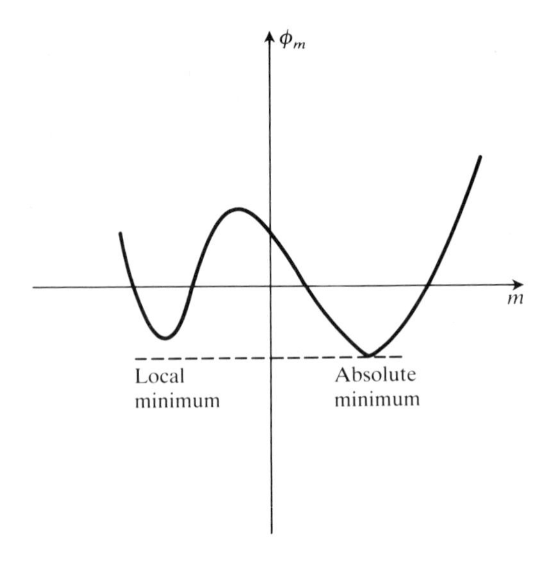

Fig. 8. Behaviour of ϕ_m for $T < T_c$ and $h > 0$.

spins and so is unaffected by local modifications of state. On the other hand, it is not globally stable, since $m^{(0)}$ is not an absolute minimum for ϕ_m. Thus, $\phi_{m^{(0)}}$ satisfies the ideal metastability conditions (IM1,2). To show that it also satisfies (IM3) we observe that, as $m^{(0)}$ is a local minimum of ϕ_m, we may choose a closed interval $I^{(0)}$, centred at $m^{(0)}$, such that the restriction of ϕ_m to $I^{(0)}$ is minimized at $m^{(0)}$. Hence, defining $\Omega^{(0)}$ to be the set of states of Σ whose polarizations are dispersion-free, and have values in the interval $I^{(0)}$, it follows that $\rho_{m^{(0)}}$ minimizes the free energy density over $\Omega^{(0)}$ and therefore satisfies (IM3). It is therefore ideally metastable.

Note. One may also prove† from (6) and (7) that the polarization $m^{(0)}(h, T)$ of this state is a smooth continuation in h, through $h = 0$, of the equilibrium polarization, $\bar{m}(-h,T)$, for the reversed field, $-h$ (cf. Fig. 9).

Note on gravitational systems. A class of models with realistic interactions that appear very likely to have ideal metastable states are the neutral systems of fermions with Newtonian gravitational and possibly also Coulomb electrostatic forces. Hertel and Thirring (1971) have proved‡ that the statistical mechanics of these systems reduces to a mean field theory, and that they undergo phase transitions of the mean field theoretic type. One therefore anticipates that they support metastable states with properties

† Cf. Emch (1967*b*)

‡ See also Hertel, Narnhofer, and Thirring (1972) for an extension of the theory to gravitational systems of charged particles, and Messer (1981) for a more refined analytical treatment of the phase transition.

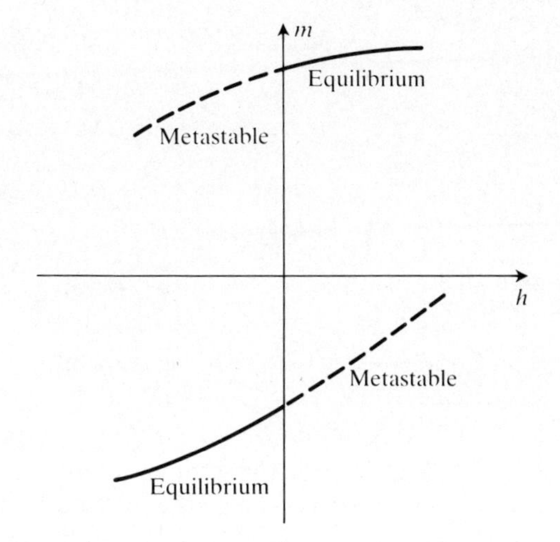

Fig. 9. Polarization as function of external field for equilibrium and metastable phases.

analogous to those obtained above in Example 2. However, there remain some difficulties† in proving this, as states of gravitational systems have rather special structures, due to the fact that the Newtonian attraction leads to collapse with volume and energy scaling as N^{-1} and $N^{7/3}$, respectively, N being the number of particles.

6.2.2. Non-translationally invariant states; surface effects

We come now to a class of metastable states whose excess energies, above the equilibrium values, correspond to surface, rather than volume, effects. These are non-translationally invariant states, consisting of different coexistent phases, separated by boundary regions, with excess free energy determined by surface tension between these phases. We shall now demonstrate, with the example of the Ising ferromagnet, that such metastable states can exist even in systems with short-range interactions.

We start by considering the Ising ferromagnet at zero temperature. There, the state, $\rho^{(+-)}$, for which all spins on one side of a plane Π are positive and those on the other side are negative has an excess surface energy over the equilibrium states, with all spins aligned. This state is nevertheless locally stable, at $T=0$, since it costs energy to reverse any finite set of spins: for the reversal of groups of spins adjacent to Π increases the area of the surface separating positive and negative spins (cf. Fig. 11), and the creation of wrongly polarized islands in the interiors of the upper and lower halves of the system manifestly costs energy. Thus, the state

† See Narnhofer and Sewell (1980) for a discussion of these difficulties, and for the formulation of states of gravitational systems. We remark that, although the states in question correspond to pure phases, they are not translationally invariant, as their densities are non-uniform.

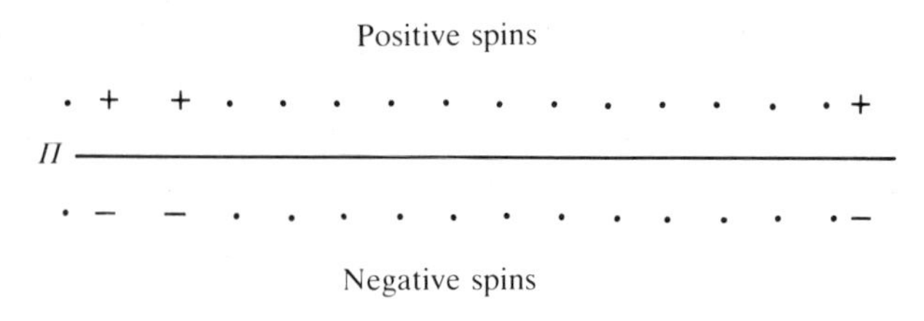

Fig. 10. Phase-separated state $\rho^{(+-)}$ of Ising system at $T=0$.

$\rho^{(+-)}$ is ideally metastable at $T=0$ on the grounds that (a) it is locally stable, (b) it has excess surface energy above that of the true ground states, with all spins aligned, and (c) it minimizes the energy density over the set, $\Omega^{(0)}$, of states whose polarizations in the regions above and below Π are respectively positive and negative. Clearly, the metastability of this state emerges only when one looks at surface effects: for, in the infinite system, it has the same energy density as the equilibrium states.

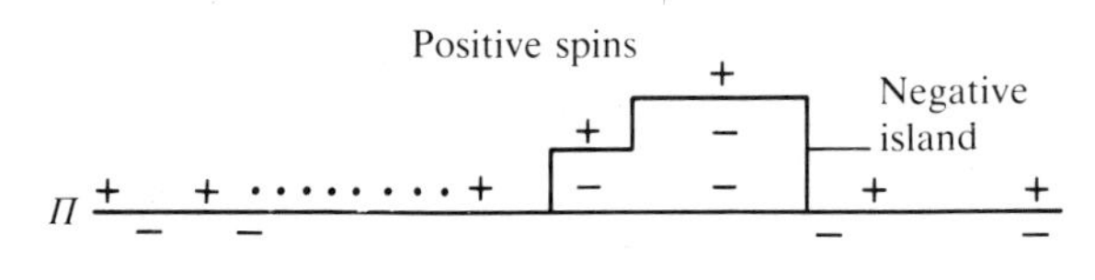

Fig. 11. Enlargement of surface of separation by reversing spins adjacent to Π.

The conclusion that the model has phase-separated ideally metastable states at $T=0$ can be extended to finite temperatures, for the *three-dimensional*† Ising ferromagnet, by virtue of results of Dobrushin (1972) and Van Beijeren (1975). For they have proved that, at sufficiently low temperatures, the infinite volume limit, $\rho_T^{(+-)}$, of the canonical state of the system, with positive boundary conditions above Π and negative ones below it, has corresponding opposite polarizations on the two sides of that surface. This state, $\rho_T^{(+-)}$, is locally stable, since it is an infinite volume limit of KMS states‡ and therefore has the KMS property.§ Hence, by the same argument used for $T=0$, the state $\rho_T^{(+-)}$ is ideally metastable.

Note. To be precise, $\rho_T^{(+-)}$ is metastable for *free boundary conditions*,

† Gallavotti (1972) and Aizenmann (1980) has proved that the two-dimensional Ising model does not support such phase separations.

‡ The KMS condition has to be formulated here in terms of the algebra of observables based on the Pauli spins $(\sigma_x, \sigma_y, \sigma_z)$, whose z-components are the Ising spins appearing in the Hamiltonian.

§ The proof that the KMS property survives the infinite volume limit is due to Haag, Hugenholtz, and Winnink (1967).

where the above consideration about its thermodynamical stability properties are applicable, even though the state was constructed as the infinite volume limit of the canonical one for $(+\ -)$ boundary conditions, positive above Π and negative below it. On the other hand, $\rho_T^{(+-)}$ is an equilibrium state for these latter conditions.

6.3. Normal metastability

We characterize normal, or non-ideal, metastable states by stability properties that give rise to 'very long', but finite, lifetimes. Such stability is evidently of a lower grade than that given by the KMS property, which leads to infinite lifetimes.

In order to formulate a definition of normal metastability, we first note that the concept of a 'very long lifetime' must depend on (a) the observational time-scale, (b) the quantities to be measured, and (c) the dynamics of the system when coupled to some thermal reservoir. All of these factors are governed by the experimental set-up. We shall generally assume that there is some given characteristic observational time, τ_0, a set $\mathscr{B} = \{B\}$ of observables, designated as the quantities to be measured, and an evolution $\rho \to \rho_t$ of the states satisfying the principle of detailed balance, stemming from the KMS property of the reservoir.‡ Under these assumptions, we term the state ρ 'very long-lived' if the changes, $|\rho_t(B) - \rho(B)|$, in the expectation values of $\mathscr{B}$-class observables are negligible over times, t, of the order of τ_0. Thus, we say that ρ has a very long lifetime, τ (at least), if

$$|\rho_t(B) - \rho(B)| < g(t/\tau)\ \|B\| \text{ for all } B \text{ in } \mathscr{B}, \tag{8}$$

where $\tau \gg \tau_0$ and g is a function of t/τ such that

$$g(s) \to 0 \quad \text{as} \quad s \to 0. \tag{9}$$

We then define a state ρ to be *normally stable* if it satisfies the following conditions, obtained by weakening the local stability requirement, that appeared in the definition of ideal metastability, to one of a 'very long lifetime'.

(NM1) *ρ is neither an equilibrium nor an ideally metastable state.*

(NM2) *ρ minimizes the free energy density over a reduced state space, $\Omega^{(0)}$, that is closed with respect to space translations and contains the limits of its sequences of translationally invariant states.*

(NM3) *ρ has a 'very long lifetime', in the sense that the conditions* (8) *and* (9) *are fulfilled for the evolution $\rho \to \rho_t$, governed by the interaction of the system with the thermal environment and so satisfying the condition of detailed balance.*

‡ Cf. §3.7.

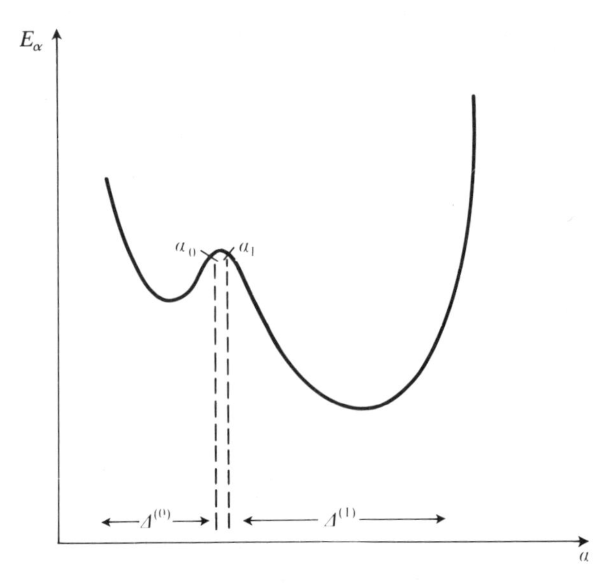

Fig. 12. Energy profile for double well model.

In view of this definition, the basic problem of normal metastability is to characterize the features of the reduced state space $\Omega^{(0)}$ that permit only very slow escape from it and thus lead to very long lifetimes for the state(s) defined by (NM2). A natural thing to look for is an energy barrier ΔE ($\gg kT$) that obstructs the escape from $\Omega^{(0)}$ and so leads to a lifetime

$$\tau \sim \tau_m \exp(\Delta E/KT), \tag{10}$$

where τ_m is some microscopic time. In this case, τ could be enormously long, because of its exponential dependence on $\Delta E/kT$ ($\gg 1$). Our aim now is to substantiate this picture of metastability, due to energy barriers, by model examples. We shall start with a systematic treatment of the dynamics of the escape from an energy well in a model whose energy spectrum has a double well structure (cf. Fig. 12). In particular, we shall show that it has metastable states based on the well of higher energy. We shall then show how an analogous treatment of the Ising ferromagnet establishes the metastability of states polarized in opposition to a weak external field: there the barrier against spin reversals is provided by the polarization field, created collectively by the spins.

6.3.1. The double well model

We take the model, Σ, to be a finite system, whose eigenstates and corresponding energy levels are $\{\phi_\alpha\}$ and $\{E_\alpha\}$, respectively, with α running over a finite set of real numbers. It is assumed that the form of E_α, as a function of the discrete variable α, corresponds to a double well (cf.

Fig. 12). Specifically, we assume that the range of values of α consists of two subranges $\Delta^{(0)}$ $(\alpha \leqslant \alpha_0)$ and $\Delta^{(1)}$ $(\alpha \geqslant \alpha_1)$, that the form of E_α over each of these is well-shaped, and that the well based on the $\Delta^{(0)}$-subrange has the higher minimum, $\alpha^{(0)}_{\min}$.

We assume that the system is coupled to a thermal reservoir, and denote by ρ_t the state (density matrix) that evolves in time t from an initial one, ρ, under the influence of this coupling. We denote the occupation probability of the eigenstate ϕ_α for the state ρ_t by $P_t(\alpha)$ $(\equiv(\phi_\alpha, \rho_t\phi_\alpha))$ and assume that P_t evolves according to a Pauli master equation

$$\frac{\mathrm{d}P_t(\alpha)}{\mathrm{d}t} = \sum_{\alpha' \neq \alpha} (W(\alpha, \alpha')P_t(\alpha') - W(\alpha', \alpha)P_t(\alpha)), \tag{11}$$

where the transition rates W satisfy the detailed balance condition

$$W(\alpha, \alpha')\exp(-\beta E_{\alpha'}) = W(\alpha', \alpha)\exp(-\beta E_\alpha), \tag{12}$$

β being the inverse temperature of the reservoir. We assume that the transitions between the two wells take place exclusively via the eigenstates ϕ_{α_0} and ϕ_{α_1} corresponding to the extremities α_0 and α_1 of the α-subranges $\Delta^{(0)}$ and $\Delta^{(1)}$. Thus,

$$\left.\begin{aligned} &W(\alpha_0, \alpha_1) = \nu \neq 0, \\ \text{and}\quad &W(\alpha, \alpha') = 0 \text{ for } \alpha \in \Delta^{(0)} \text{ and } \alpha' \in \Delta^{(1)} \\ &\qquad\qquad \text{unless } \alpha = \alpha_0 \text{ and } \alpha' = \alpha_1. \end{aligned}\right\} \tag{13}$$

We denote by ΔE the difference between the energy of ϕ_{α_0} and the minimum energy, $E^{(0)}_{\min}$, for the $\Delta^{(0)}$-well, i.e.

$$\Delta E = E_{\alpha_0} - E^{(0)}_{\min}. \tag{14}$$

Thus, ΔE represents an energy barrier obstructing the escape from this well by the only available channel (i.e. via ϕ_{α_0}). It will be assumed that

$$\Delta E \gg kT. \tag{15}$$

Our aim now is to show that the model has metastable states in the $\Delta^{(0)}$-well with lifetimes of the form (10). Thus, we define $\Omega^{(0)}$ to be the set of states of Σ based on eigenstates in that well, i.e. $\Omega^{(0)}$ consists of the states ρ such that $\rho\phi_\alpha = 0$ for α in $\Delta^{(1)}$. We define $\rho^{(0)}$ to be the canonical state, subject to the system being constrained to the reduced state space $\Omega^{(0)}$, i.e.

$$(\phi_{\alpha'}, \rho^{(0)}\rho_\alpha) = P^{(0)}(\alpha)\delta_{\alpha\alpha'}, \tag{16}$$

where

$$P^{(0)}(\alpha) = \begin{cases} \exp(-\beta E_\alpha) \Big/ \sum_{\alpha \in \Delta^{(0)}} (\text{idem}) & \text{if} \quad \alpha \in \Delta^{(0)} \\ 0 & \text{if} \quad \alpha \in \Delta^{(1)}. \end{cases} \tag{17}$$

We assume that $\mathscr{B}$, the set of quantities to be measured, consists of the observables, $\{B\}$, whose eigenstates are $\{\phi_\alpha\}$: the corresponding eigenvalues of B will be denoted by $\{B_\alpha\}$. The time-dependent expectation values of these observables for an *arbitrary* time-dependent state ρ_t is then given by the formula

$$\rho_t(B) = \sum_t P_t(\alpha) B_\alpha, \tag{18}$$

where, as above, $P_t(\alpha)$ is the occupation probability of ϕ_α and evolves according to the master equation (11).

It follows easily from our definitions of $\Omega^{(0)}$ and $\rho^{(0)}$ that this state satisfies the metastability conditions (NM1,2).† To prove that $\rho^{(0)}$ satisfies (NM3), we need to analyse the time-development of its associated probability distribution, $P_t^{(0)}(\alpha)$, as governed by the master equation (11). For this we first note that, in view of the linearity of that equation, its general solution takes the form

$$P_t(\alpha) = \sum_{\alpha'} T_t(\alpha, \alpha') P_0(\alpha'), \tag{19}$$

where $T_t(\alpha, \alpha')$ is the solution corresponding to the initial condition $P_0(\alpha) = \delta_{\alpha\alpha'}$, and is therefore the transition probability from $\phi_{\alpha'}$ to ϕ_α in time t. Hence it possesses the positivity and normalization properties

$$T_t(\alpha, \alpha') \geqslant 0 \quad \text{and} \quad \sum_\alpha T_t(\alpha, \alpha') = 1. \tag{20}$$

Further, as the master equation is invariant under time-translations $t \to t + \text{const.}$, its solution also relates P_{t+s} to P_s, for $s > 0$, by the following analogue of (19):

$$P_{t+s}(\alpha) = \sum_{\alpha'} T_t(\alpha, \alpha') P_s(\alpha'). \tag{21}$$

Therefore, as $\mathrm{d}P_t/\mathrm{d}t \equiv (\mathrm{d}P_{t+s}/\mathrm{d}s)_{s=0}$, it follows from (11) and (21) that this time-derivative may be expressed in terms of the initial probability distribution, P_0, the transition rates W and the transition probabilities T_t by the formula

$$\frac{\mathrm{d}P_t(\alpha)}{\mathrm{d}t} = \sum_{\alpha', \alpha''} T_t(\alpha, \alpha')(W(\alpha', \alpha'')P_0(\alpha'') - W(\alpha'', \alpha')P_0(\alpha')), \tag{22}$$

and correspondingly, by (18),

$$\frac{\mathrm{d}}{\mathrm{d}t}\rho_t(B) = \sum_{\alpha, \alpha', \alpha''} T_t(\alpha, \alpha') B_\alpha (W(\alpha', \alpha'')P_0(\alpha'') - W(\alpha'', \alpha')P_0(\alpha')). \tag{23}$$

In particular, if ρ is the state $\rho^{(0)}$ and thus, by (16), the initial probability is

† The translational invariance appearing in (NM2) is evidently irrelevant here. Also, the possibility of ideal metastability does not arise, as the system is finite.

$P^{(0)}$, then

$$\frac{\mathrm{d}}{\mathrm{d}t}\rho_t^{(0)}(B) = \sum_{\alpha,\alpha',\alpha''} T_t(\alpha,\alpha')B_\alpha[W(\alpha',\alpha'')P^{(0)}(\alpha'') - W(\alpha'',\alpha')P^{(0)}(\alpha')]. \tag{24}$$

By the definition (17) of $P^{(0)}$ and the detailed balance condition (12), the expression in square brackets in this last equation vanishes if α', α'' are both in $\Delta^{(0)}$ or both in $\Delta^{(1)}$. Hence, by (13), the only pairs of values of α' and α'' for which this expression does not vanish are $\alpha' = \alpha_0$ and $\alpha'' = \alpha_1$, and $\alpha'' = \alpha_0$ and $\alpha' = \alpha_1$. The corresponding values of the expression in square brackets are then $\nu P^{(0)}(\alpha_0)$ and $-\nu P^{(0)}(\alpha_0)$, respectively, and so eqn (24) reduces to the form

$$\frac{\mathrm{d}}{\mathrm{d}t}\rho_t^{(0)}(B) = \nu \sum_\alpha (T_t(\alpha,\alpha_1) - T_t(\alpha,\alpha_0))B_\alpha P^{(0)}(\alpha_0). \tag{25}$$

Therefore, as $|B_\alpha| \leqslant \|B\|$ and T_t satisfies the conditions (20),

$$\left|\frac{\mathrm{d}}{\mathrm{d}t}\rho_t^{(0)}(B)\right| \leqslant 2\nu \, \|B\| \, P^{(0)}(\alpha_0)$$

$$\leqslant 2\nu \exp(-\Delta E/kT), \text{ by (14) and (17).}$$

Hence,

$$|\rho_t^{(0)}(B) - \rho^{(0)}(B)| \leqslant 2\nu t \exp(-\Delta E/kT) \, \|B\|, \tag{26}$$

and therefore, as $\Delta E \gg kT$ (by (15)), it follows that $\rho^{(0)}$ satisfies the metastability condition (NM3) and has a 'very long' lifetime

$$\tau = (2\nu)^{-1} \exp(\Delta E/kT). \tag{27}$$

The state $\rho^{(0)}$ is therefore normally stable, and owes its long lifetime to the energy barrier ΔE ($\gg kT$) obstructing escape from the $\Delta^{(0)}$-well.

6.3.2. *States of an Ising ferromagnet with polarization opposed to an external field*

We consider here a two-dimensional Ising ferromagnet on a square lattice. The Hamiltonian for the model, in the presence of an external field, h, is given formally by the expression

$$-J\sum_{x,x'}{}' \sigma(x)\sigma(x') - h\sum_x \sigma(x), \tag{28}$$

where, in a usual way, $J > 0$, $\sigma(x)$ takes values ± 1, and the prime over the first sigma indicates that summation is confined to nearest neighbours. It will be assumed that h is much smaller, in absolute magnitude, than J, which represents a measure of the 'inner field' due to the polarization of the system in the condensed phase. It is the domination of the external field by the internal one that gives rise to the possibility of metastable states with polarizations opposed to the applied field.

In order to see the nature of the competition between the internal and external fields, we consider first the stability properties of the state of the system in which all the spins are aligned in opposition to the applied field. The energy required to reverse a single spin, when the system is in that state, is evidently $2(4J - |h|)$, which is positive as $J \gg |h|$. Likewise, the energy required to reverse any 'small' number ($\ll J/|h|$) of spins is positive. The situation changes, however, when one considers the reversal of all the spins in a 'large' square Δ_L whose side is of length L, in units of the lattice spacing. For the resultant energy increment is then

$$\Delta E_L = 2(4JL - |h|\, L^2), \tag{29}$$

the first term representing the 'surface effect' of the interaction of the spins at the boundary of Δ_L with their neighbours outside that square, and the second term representing the bulk effect of their interaction with the external field. The energy ΔE_L obtains a maximum value

$$\Delta E_{\max} \simeq 8J^2/|h| \tag{30}$$

for $L \simeq 2J/|h|$ and becomes negative only when $L \gtrsim 4J/|h|$. Thus, the 'wrongly polarized' state is stabilized against spin reversals by energy barriers $\simeq 8J^2/|h|$, corresponding to the formation of 'droplets' of $\simeq (2J/|h|)^2$ spins aligned to the field. The same argument may also be applied to 'wrongly polarized' states at finite temperature, T, below the critical point, T_c. For suppose that the system is prepared in an equilibrium state $\rho_{T,-h}$ at temperature T ($<T_c$) in the presence of an external field $-h$, and that the field is then subsequently reversed. $\rho_{T,-h}$ is then no longer an equilibrium state, as its polarization opposes the external field, h. Furthermore, if the spins occupying a square of side L are reversed when the system is in this state, the entropy change will be zero,† while the increment in energy, and thus in free energy, will be of the same form as (29), but with the coefficients of L and L^2 modified by factors, of the order of unity, whose precise values depend on T and h. Hence, arguing as previously, one finds that the wrongly polarized state, $\rho_{T,-h}$, is stabilized against spin reversals by energy barriers, of the order of $J^2/|h|$, corresponding to the formation of 'rightly polarized droplets' of $\sim J^2/|h|^2$ spins. Since $J \sim kT_c \gg |h|$ and $T < T_c$, these barriers $\gg kT$. This suggests that the state $\rho_{T,-h}$ should have a very long lifetime and so should be metastable, by the same mechanism that operated in the double well model. We shall now sketch a rigorous treatment, due to Capocaccia, Cassandro, and Olivieri (1974), that has established the metastability of wrongly polarized states of the *finite* version of the model by essentially that mechanism. As we shall subsequently

† This follows from the definition of incremental entropy in §3.4.3.

indicate, the situation is less clear for the infinite version of the model, and it appears that it is the finite one that captures the essential features of normal metastability.

The finite system. The finite version of the model is taken to be a system Σ_Λ of Ising spins occupying a square, Λ, with free boundary condition, in the presence of a weak, negative external field, h. The Hamiltonian is thus given by (28), with summations confined to Λ. Denoting an arbitrary configuration of spins $\{\sigma(x) \mid x \in \Lambda\}$ by $\boldsymbol{\sigma}$, we take the states to be the probability distributions, $\{P(\boldsymbol{\sigma})\}$, of $\boldsymbol{\sigma}$. It is assumed that the system interacts with a reservoir by a coupling that leads to single spin reversals. Thus, denoting by $\boldsymbol{\sigma}'_x$ the spin configuration obtained from $\boldsymbol{\sigma}$ by reversing the spin at x, it is assumed that the time-dependent state $P_t(\boldsymbol{\sigma})$ evolves according to a master equation†

$$\frac{\mathrm{d}P_t(\boldsymbol{\sigma})}{\mathrm{d}t} = \sum_{x \in \Lambda} (W(\boldsymbol{\sigma}, \boldsymbol{\sigma}'_x) P_t(\boldsymbol{\sigma}'_x) - W(\boldsymbol{\sigma}'_x, \boldsymbol{\sigma}) P_t(\boldsymbol{\sigma})), \tag{31}$$

where W satisfies the detailed balance condition,

$$W(\boldsymbol{\sigma}, \boldsymbol{\sigma}') = W(\boldsymbol{\sigma}', \boldsymbol{\sigma}) \exp(\beta \Delta E(\boldsymbol{\sigma}, \boldsymbol{\sigma}')), \tag{32}$$

$\Delta E(\boldsymbol{\sigma}, \boldsymbol{\sigma}')$ being the increment in energy for a transition from $\boldsymbol{\sigma}$ to $\boldsymbol{\sigma}'$. It is also assumed that $W(\boldsymbol{\sigma}, \boldsymbol{\sigma}'_x)$ has strictly positive upper and lower bounds, independent of Λ, x, and $\boldsymbol{\sigma}$, i.e.

$$W_M \geqslant W(\boldsymbol{\sigma}, \boldsymbol{\sigma}'_x) \geqslant W_m > 0. \tag{33}$$

The set, $\mathscr{B}$, of quantities to be measured is taken to consist of all the functions, $B(\boldsymbol{\sigma})$, of the spin configurations. The expectation value of these observables for the state P_t are given by

$$P_t(B) = \sum_{\boldsymbol{\sigma}} P_t(\boldsymbol{\sigma}) B(\boldsymbol{\sigma}). \tag{34}$$

The states constructed by Capocaccia, Cassandro, and Olivieri (CCO) (1974) as wrongly polarized metastable states of Σ_Λ are not precisely the equilibrium states, $P_{T,-h}$, for the reversed field, $-h$, but close approximants to these. Their construction was based on the observation that the probability for a region bounded by a 'curve' γ to be negatively polarized, when the state is $P_{T,-h}$, decreases exponentially with $J|\gamma|/kT$, $|\gamma|$ being the perimeter of γ; and consequently this state may be approximated arbitrarily closely by ones that admit no islands of negative spins that exceed some suitably chosen size. Accordingly, they introduced a reduced set, $\Delta^{(0)}$, of spin configurations, defined by the following conditions.

(a) The spins inside and adjacent to the boundary of Λ are all positive. This

† This particular master equation is due to Glauber (1963*b*).

implies that each spin configuration, $\boldsymbol{\sigma}$, corresponds to a unique set of polygonal Peierls contours† separating positive from negative spins, and that the negative ones are all enclosed within these contours.

(b) The areas (i.e. the number of sites) enclosed by these contours do not exceed a certain critical value, L_0^2, chosen to represent essentially the size ($\sim (J/|h|)^2$) that a negative droplet must reach before the energy barrier, created by the positive polarization, can be overcome. Thus,

$$L_0 = \alpha J/|h| \tag{35}$$

where α is a parameter of the order of unity. As we shall presently see, there is some latitude in the precise choice of α. The reduced state space $\Omega^{(0)}$, based on $\Delta^{(0)}$, consists of those states for which the spin configuration lies in $\Delta^{(0)}$, with probability equal to 1. The canonical state of Σ_Λ, in the field h, subject to the constraint to this reduced state space, is given by the probability distribution

$$P^{(0)}(\boldsymbol{\sigma}) = \begin{cases} A\exp(-\beta E(\boldsymbol{\sigma})) & \text{for } \boldsymbol{\sigma} \text{ in } \Delta^{(0)} \\ 0 & \text{otherwise} \end{cases} \tag{36}$$

where $E(\boldsymbol{\sigma})$ is the energy for the configuration $\boldsymbol{\sigma}$ and A is the normalization factor. CCO obtained bounds on the correlation functions for this state, on the basis of Peierls contour estimates, and thereby proved that $P^{(0)}$ enjoyed the following properties, (c) and (d), subject to the conditions that‡

$$kT < J/\ln 6 \tag{37}$$

and

$$\alpha(\equiv L_0|h|/J) < 4(1 - kT\ln 6/J). \tag{38}$$

(c) $P^{(0)}$ reduces, in the limit where first $\Lambda \to \infty$ and then $h \to -0$, to the equilibrium state corresponding to a positively polarized phase in the absence of any applied field. In other words, for 'large Λ and small negative h', $P^{(0)}$ simulates the true equilibrium state of the system for the reversed field, $-h\ (>0)$.

(d) The probability that $\gamma_1, \ldots, \gamma_k$ are the outer polygons, enclosing islands of negative spins, is majorised, for the state $P^{(0)}$, by $\exp \times (-a\sum_1^k |\gamma_j|)$, where $a \geqslant 2\ln 6$ and $|\gamma|$ is the length of γ. Hence, one may use the Peierls argument to prove that the system is positively polarized in this state. For it follows from the form of the upper bound, for the contour probability distribution, that we have just cited, that the probability of finding a negative spin at an arbitrary site x is majorized

† See §3.5.1 for specifications of these contours.
‡ These are merely conditions that have been proved to be sufficient for (c) and (d).

by $\Sigma_4^\infty N(l)e^{-al} \leqslant \Sigma_4^\infty N(l)6^{-2l}$, where $N(l)$ is the number of different contours of length l that can be drawn to surround x. Since, as proved in §3.5.1, $N(l) < l^2 . 4^{(l+1)}$, it is a simple matter to check that this probability $< \frac{1}{2}$ and that, consequently, the state $P^{(0)}$ is positively polarized.

Thus, under the conditions (37) and (38), $P^{(0)}$ is a wrongly polarized state that simulates one of thermal equilibrium in the reversed field, $-h$. We now consider the candidacy of $P^{(0)}$ for metastability. First, we remark that the condition (38) permits some latitude in the precise value of L_0 and thus in the form of $P^{(0)}$. However, the specific choice of L_0 is unimportant since, by (d), the dependence of $P^{(0)}$ on this parameter is negligible. The central issue, then, is whether, for a given L_0 that satisfies (38), the state $P^{(0)}$ possesses the metastability properties (NM1–3). In fact, it has been established by CCO that it does so, on the basis of an argument analogous to the one we employed in §6.3.1 for the double well model. Let us now outline how the argument proceeds for the Ising system.

It follows easily from our definitions of $\Omega^{(0)}$ and $P^{(0)}$ that this state satisfies the conditions (NM1,2). The question of whether it also satisfies (NM3) evidently depends on the evolution of the system, according to the master equation (31), when $P^{(0)}$ is the initial state. Since this evolution satisfies the condition of detailed balance and since $P^{(0)}$ is the canonical state for the reduced state space $\Omega^{(0)}$, we have a situation similar to that formulated in §6.3 for the double well model. In particular, the time-development of $P_t^{(0)}$ is achieved, both for the Ising system and the double well model, through the escape from $\Omega^{(0)}$. In the case of the latter model, that amounts to the escape from one well and requires an activation energy ΔE corresponding to the height of the relevant energy barrier. In the case of the Ising system, it amounts to the growth of negatively polarized 'droplets' beyond the critical size L_0^2. Since this can only be achieved by reversals of spins adjacent to droplets of this critical size, the relevant activation energy is that required for the formation of such droplets, i.e. $\sim J^2/|h|$. The one important respect in which the kinetics of the Ising system differs from that of the double well model, is that whereas this latter model has only one channel of escape from $\Omega^{(0)}$, namely by transitions from the eigenstate ϕ_{α_0} to ϕ_{α_1} (cf. §6.3.1), the Ising system has many channels of escape, corresponding to the various positionings of the critical droplets and of the sites adjacent to them: thus, the number of such channels is proportional to $|\Lambda|$.

These elementary considerations of the kinetics of the Ising system would lead one, then, to anticipate that the lifetime of $P^{(0)}$ is of the same form as given by (27) for that of the metastable state of the double well system, with ΔE replaced by an energy $\sim J^2/|h|$, and with an additional factor $\sim |\Lambda|^{-1}$ to take account of the various positionings of the critical droplets. This anticipation is, in fact, borne out by the rigorous treatment, by CCO, of the master equation (31). Their essential result is that the state $P^{(0)}$ satisfies the metastability condition (IM3), provided that $J^2/|h|\,kT \gg \ln|\Lambda|$, and that its lifetime is not less than

$$\tau = \frac{a}{W_M\,|\Lambda|}\exp(bJ^2/|h|\,kT), \tag{39}$$

where a (~ 100) and b (~ 1) are numerical factors and W_M is the maximum spin reversal frequency occurring in the master equation. The wrongly polarized state $P^{(0)}$ is therefore metastable if $J^2/|h|\,kT \gg \ln|\Lambda|$, a condition that can easily be satisfied for realistic values of the parameters of the model, e.g. $|\Lambda| \sim 10^{16}$, $J/kT \sim 3$ and $J/|h|$, the ratio of the polarization to the external field, ~ 100.

The infinite system. Since the lifetime, τ, as given by (39), is inversely proportional to $|\Lambda|$, it would appear that wrongly polarized metastable states can occur only in the finite (and not too large) version of the model. That conclusion, however, would be based on the assumption that the set of relevant observables consisted of all the functions of the spins. A different picture, which has been explored by Vanheuverszwijn (1979), emerges if $\mathscr{B}$ is chosen to consist only of intensive macroscopic variables – an appropriate choice if one is concerned only with thermodynamical observations. For the rates of change of these variables (densities of extensive quantities) are governed by the rate of formation *per unit volume* of the critical droplets, and so the lifetimes of the wrongly polarized states may still be long in the macroscopic picture based on these variables.

The treatment and results of Vanheuverszwijn may be briefly summarized as follows. The reduced state space $\Omega^{(0)}$ for the infinite system is defined analogously with that for the finite one, $P^{(0)}$ is taken to be the state that minimized the free energy density over $\Omega^{(0)}$, the dynamics is again assumed to be given by the master equation (31), and $\mathscr{B}$ is taken to consist of just the spin density, i.e. polarization of the system. It follows from these specifications that $P^{(0)}$ is a wrongly polarized state satisfying the conditions (NM1, 2).† To formulate (NM3), Vanheuverszwijn noted that it followed from translational invariance that the time-dependent polarization m_t, for an evolution from the state $P^{(0)}$, was just the expectation value, at time t, of the spin at the origin, O, and obtained the following upper bound for the modulus of the increment $(m_t - m_0)$:

$$|m_t - m_0| < aW_M t\exp(4W_M t)\exp(-bJ^2/|h|\,kT), \tag{40}$$

where a and b are the same parameters as in (39), and the factor $\exp(4W_M t)$ represents an *overestimate* of the enhancement of the probability of reversal

† Here, ideal metastability is ruled out because of the short-range forces.

of the spin at O, due to the growth of negatively polarized droplets, that attain critical size in regions that exclude the origin and then spread so as to cover that point. It follows from (40) that, if $J^2/|h|\,kT \gg 1$, then the change in polarization is negligibly small until t reaches a value $\simeq \bar{\tau}$, where

$$\bar{\tau} = \frac{b}{4W_M} J^2/|h|\,kT. \tag{41}$$

Since this time tends to infinity as $h \to -0$, it follows that the metastability condition (NM3) is fulfilled for sufficiently small $|h|$. However, the external field would really have to be extraordinarily weak for this condition to be satisfied. For example, if W_M were a typical microscopic spin reversal frequency (e.g. $10^{-8}\,\text{sec}^{-1}$) and if $kT \sim J/3$, then $J/|h|$ would have to be of the order of 10^{10}, as against $\sim 10^6$ for the weakest detectable fields, if $\bar{\tau}$ were to be one hour.

This does not mean, of course, that the infinite Ising system cannot support wrongly polarized metastable states in fields of reasonable strength (e.g. the Earth's), since the formula (40) represents only an upper bound for the polarization change. To give an indication of why one might expect much larger estimates for the lifetime of $P^{(0)}$ to be valid, we make the simple assumption that once a droplet has reached critical size, its 'radius' grows at a rate $\sim v$, given by W_M times the lattice spacing. Under this assumption, the factor $(\exp 4W_M t)$ in (41) may be replaced by a quantity, $\sim (vt)^2$, representing the area of the region from where droplets can spread to reach 0 within time t. In that case, the lifetime is at least

$$\bar{\tau}' \sim (aW_M v^2)^{-\frac{1}{3}} \exp(bJ^2/3|h|\,kT), \tag{42}$$

which would be astronomically long, even for reasonably strong fields, e.g. those for which $J/|h| \sim 100$.

Comment on finite versus infinite systems. We now remark that, even if one grants the validity of the heuristic argument leading to (42), it cannot represent the mechanism by which a wrongly polarized state of a real magnet is metastable unless R, the length span of the magnet, exceeds $\sim v\bar{\tau}'$, the radius of the circle from which droplets can spread to cover O within the lifetime of the state. In fact, it is easy to see that this condition would not be satisfied for reasonable values of the parameters if the magnet were a typical laboratory specimen, with $R \sim 1$–100 cm. For, if $\bar{\tau}'$ were the observed lifetime, then even taking a conservative estimate of $\bar{\tau}'$ as one hour and v, the product of the spin reversal frequency W_M and the lattice spacing, as 10 cm/sec, the length $v\bar{\tau}'$ would be of the order of kilometres.

Furthermore, it follows from the formulae (39) and (42) for the lifetimes of the finite and infinite versions of the model that, as $v = W_M$ in units where the lattice spacing is unity,

$$\bar{\tau}' < \tau \quad \text{if} \quad a^{\frac{4}{3}} |\Lambda|^{-1} \exp(\tfrac{2}{3} bJ^2/|h|\,kT) \gtrsim 1 \tag{43}$$

Therefore, since even τ represents an underestimate for the lifetime of a magnet occupying the region Λ, it follows that this provides a better estimate than $\bar{\tau}'$ for that lifetime if the condition given by (43) is fulfilled; and one sees that it is when the parameters of the model take the typical values given after eqn (39). Thus, in such cases, it appears that the finite, rather than the infinite, model properly represents the essential features of metastability.

6.4. Concluding remarks

The picture of metastability that emerges from the theory presented here is that it corresponds to stabilization of states by energy barriers, that are created by the collective action of the macroscopic components of a system and that impede its escape into states of lower free energy. In other words, the metastable states are *self-stabilized*. This is exemplified by the wrongly polarized states of ferromagnets, which are stabilized against spin reversals by their own 'inner fields'.

The essential difference between ideal and normal metastability may be described in these terms by saying that, in the ideal case, the stabilizing barriers cannot be overcome by any localized modifications of state, whereas in the normal case they can, although that requires an activation energy much larger than kT. This difference is reflected by the mathematical structures of the theories of the two types of metastability. For the theory of ideal metastable states involves no explicit kinetic theory, as the local stability of these states guarantees that they have infinite lifetimes: the theory of normal metastable states, on the other hand, has an essential kinetic theoretic component, that determines their lifetimes. Furthermore, whereas the theory of normal metastability appears to be most naturally based on the model of 'large, but finite' systems,† the precise formulation of ideal metastability can only be achieved within the framework of the statistical mechanics of infinite systems, since only here is the distinction between global and local stability clear-cut.

We may summarize the developments discussed in this chapter by saying that the present scheme, based on the classification of metastable states as normal or ideal, provides a framework for a theory of metastability. On the other hand, rigorous treatments of models of metastability are very scanty, being confined essentially to the rudimentary models of the type discussed here. Evidently, a systematic treatment of a wide class of models is needed to provide insights into the vast variety of metastable states that exist in nature.

† Cf. discussion at the end of §6.3.2.

7

Order and phase transitions far from equilibrium

7.1. Introduction

Ordered structures, characterized by spontaneous symmetry breakdown or long-range correlations, are by no means confined to thermal equilibrium states. For such structures are common in open macroscopic systems far from equilibrium in such varied contexts as laser physics, hydrodynamics, chemistry, and biology. Let us briefly describe some of them from an empirical standpoint, before discussing the underlying theory.

(a)† The light emitted by a laser becomes coherent, in the sense of being monochromatic with sharply defined phase angle, when the intensity of the (random) optical pumping, that sustains it, exceeds a certain critical value. By contrast, the phase angle of a normal light beam undergoes fluctuations that destroy correlations between the electric field vector at different space–time points. The coherence of the laser light represents a breakdown of gauge symmetry, corresponding to the invariance of the dynamics of the entire apparatus under changes of the phase angle.

(b) A fluid, heated from below, takes up stationary states, with a cellular structure, when the temperature gradient is sufficiently large. This is the Bénard effect,‡ and corresponds to an order–disorder transition, in which the ordered phase has a cellular structure and thus breaks the space-transitional symmetry of the system.

(c)§ Various chemical reactions produce spatially and temporally periodic structures when the relative concentrations of the reactants lie within certain ranges.

From a theoretical standpoint, there are some clear similarities and differences between the processes that generate order in equilibrium and non-equilibrium states. To get a preliminary idea of these let us first recall that a system, Σ, may be prepared in an equilibrium state by coupling it to a thermal reservoir. For the KMS property of the reservoir induces a

† See Haken (1970).
‡ See the books by Chandrasekar (1961) and Glansdorff and Prigogine (1971) for treatments of this and other hydrodynamical phenomena whose fluids take up ordered structures.
§ See Part III of the book by Glansdorff and Prigogine (1971).

dynamics in Σ that conforms to the principle of detailed balance, with the result that this system is driven irreversibly into an equilibrium state at the reservoir temperature,† T. Furthermore, under suitable conditions on the microscopic constitution of Σ, the competition between the requirements of low energy and high entropy leads to an order–disorder transition at some critical temperature T_c. The key factors governing the equilibrium ordering are therefore the KMS property of the reservoir and the consequent energy–entropy balance in the terminal state of Σ.

In non-equilibrium situations, these two factors are not operative. Nevertheless, it is not hard to see, in specific cases, that the coupling of a system Σ to a suitable reservoir R can drive Σ irreversibly into a stable, ordered state, far from equilibrium. For example, suppose that Σ is an Ising ferromagnet and R an energy source, or pump,‡ that drives Σ into a state of maximum energy. Then assuming that the spins of Σ interact via the standard nearest neighbour coupling, $-J\sigma\sigma'$, with $J>0$, one sees immediately that its states of maximum energy are *antiferromagnetically ordered*. These are certainly not thermal equilibrium states of this system. We therefore have here a simple example of an open system that is driven to an ordered stationary state, far from equilibrium as a result of its coupling to a non-thermal reservoir. It is noteworthy that this reservoir serves to supply energy to the system and so drive it to an ordered state of low entropy whereas, by contrast, the generation of order in equilibrium states is generally achieved by withdrawal of energy from the system, i.e. by cooling.

Granted, then, that the dynamics of reservoir-driven open systems can lead to ordered non-equilibrium structures, the theoretical problem of characterizing and classifying such structures is generally more complex than for equilibrium states, because the effects of the reservoirs on the systems in question are no longer covered by any single simple condition, e.g. that of KMS. Nevertheless, there are some extremely interesting and suggestive results on non-equilibrium order–disorder transitions that have been obtained from microscopic dynamical treatments of specific models. For example, Graham and Haken (1970) have provided a statistical mechanical derivation of the transition of laser light from an incoherent to a coherent state when the optical pumping strength reaches a critical value, and have demonstrated a remarkable analogy between this and the superconducting phase transition. H. Fröhlich (1968) has constructed a kinetic theoretic model in which the combined action of thermal reservoirs and energy pumps on a system of phonons, corresponding to electrical oscillations, leads to a Bose–Einstein condensation into the mode of lowest frequency when the pumping strength is sufficiently great: this is proposed as a

† Cf. §3.7.
‡ See §7.2 for a quantum mechanical formulation of a pump.

mechanism for coherent excitations in biological systems. Thus, both this model and the laser have order–disorder transitions that are analogous to those occurring at equilibrium in other systems. This is in line with Haken's (1975) suggestion that equilibrium and non-equilibrium phase transitions generally conform to analogous laws, at an appropriate kinetic theoretic or even phenomenological level.

The object of the present chapter is to illustrate, by explicit treatments of the laser and pumped phonon models, how ordering and phase transitions arise as a result of irreversible processes in systems far from equilibrium. We shall start, in §7.2, by formulating simple open systems consisting of single atoms or oscillators coupled to thermal reservoirs corresponding to energy sources (pumps) and sinks, and demonstrating how these systems are irreversibly excited or damped down by the action of the reservoirs. We shall then, in §7.3, pass on to an account of the rigorous treatment, by Hepp and Lieb (1973), of the laser phase transition on the basis of a relatively simple model of atoms, radiation and reservoirs, due to Dicke (1954): the simple open systems of §7.2 are, in fact, the building blocks of this model. In §7.4, we shall describe Fröhlich's kinetic theoretic model of Bose–Einstein condensation in a pumped phonon system. We shall conclude, in §7.5, with a brief discussion of the problem of a general characterization of non-equilibrium order and phase transitions. The Appendix is devoted to some calculations needed for the theory of §§7.2 and 7.3.

Throughout this chapter, we employ units where $\hbar = 1$.

7.2. Simple models of open systems

We consider two simple models of open systems. These particular models are of interest both because they provide simple illustrations of how an open system, i.e. a part of a closed conservative one, can evolve according to an *irreversible* dynamic law; and because they form the building blocks of Dicke's laser model, in the version due to Hepp and Lieb (1973), that we shall treat in §7.3.

Each of the models considered here consists of a finite system, S, in interaction with an infinite one, R, which plays the role of a reservoir. It is assumed that S and R are prepared independently of one another, in states ρ_S and ρ_R respectively, and then coupled together at time $t = 0$. The initial state of the composite system, $(S + R)$, is then $\rho_S \otimes \rho_R$. Hence, assuming that the conservative dynamics of this composite system corresponds to transformations $\rho \rightarrow \rho_t$ of its states, it follows that, under the specified conditions, its state at time t is $(\rho_S \otimes \rho_R)_t$. The time-dependent state $\rho_{S,t}$ of the open system S is therefore given by

$$\rho_{S,t}(A) = (\rho_S \otimes \rho_R)_t(A \otimes I_R)$$

for all observables A of S. Hence, the dynamics of S depends both on that of the whole system $(S+R)$ and on the reservoir state ρ_R.

The models, that we shall present here, provide examples where R behaves as a sink, or absorber of quanta, that damps down the motion of S (§7.2.1); and where R consists of a sink and a pump, or emitter of quanta, that energizes S (§7.2.2).

7.2.1. *Damped harmonic oscillator*

We suppose here that S is a simple harmonic oscillator and that R is an infinite set of such oscillators, corresponding to normal modes of a free Bose field. The coupling between S and R is assumed to be harmonic. In a standard way, we formulate S in terms of creation and annihilation operators a^* and a, acting in an irreducible representation space of the canonical commutation reactions

$$[a^*, a] = 1. \tag{1}$$

We take R to be a one-dimensional Bose field $A(k)$, with the continuous variable k indicating the normal modes. Specifically, we assume that $A^*(k)$ and $A(k)$ are creation and annihilation operators satisfying the canonical commutation relations

$$[A^*(k), A(k')] = \delta(k-k'); \qquad [A(k), A(k')] = 0, \tag{2}$$

and acting in the Fock space $\mathscr{H}$, defined by the conditions that (a) it contains a 'vacuum vector' Ψ, such that $A(k)\Psi \equiv 0$, and (b) it is generated by application to Ψ of the polynomials in $\int \mathrm{d}k\, A^*(k)f(k)$, with $\int \mathrm{d}k\, |f(k)|^2$ finite. Thus, we are confining ourselves to situations where the state of the reservoir lies in the 'island'† given by the Fock representation.

We assume that the Hamiltonian for the composite system $(S+R)$ is

$$H = \omega a^* a + \int \mathrm{d}k\, f(k) A^*(k) A(k) + \int \mathrm{d}k\, g(k)(A^*(k)a + a^* A(k)), \tag{3}$$

with ω, $f(k)$ and $g(k)$ c-numbers. Thus, the first, second, and third terms represent the energies of the oscillator S, the reservoir R, and the S–R interaction, respectively. The systems S and R are assumed to be initially prepared, independently of one another, with S in some vector state ψ and R in the Fock vaccum Ψ. The initial state of $(S+R)$ is then $\psi \otimes \Psi$. Here we note that by choosing Ψ as the initial state of R, we are ensuring that this reservoir can only absorb quanta, thus acting as a perfect sink.

We formulate the dynamics of the model in Heisenberg representation. Thus, denoting the Heisenberg operators representing the time-translates of a and $A(k)$ by $a(t)\,(\equiv \mathrm{e}^{\mathrm{i}Ht} a \mathrm{e}^{-\mathrm{i}Ht})$ and $A(k,t)$ $(\equiv \mathrm{e}^{\mathrm{i}Ht} A(k) \mathrm{e}^{-\mathrm{i}Ht})$, respec-

† We recall from Chapter 2 that this island is the set of states corresponding to vectors or density matrices in the Fock representation space.

tively, we infer from the formula (3) for H and the commutation relations (1) and (2) that the equations of motion for $a(t)$ and $A(k,t)$ are the following.

$$\frac{\mathrm{d}a(t)}{\mathrm{d}t} = -\mathrm{i}\omega a(t) - \mathrm{i}\int \mathrm{d}k\, g(k)A(k,t), \tag{4}$$

and

$$\frac{\mathrm{d}A(k,t)}{\mathrm{d}t} = -\mathrm{i}f(k)A(k,t) - \mathrm{i}g(k)a(t). \tag{5}$$

On solving this last equation for A in terms of a, we obtain the formula

$$A(k,t) = A(k)\exp(-\mathrm{i}f(k)t) - \mathrm{i}g(k)\int_0^t \mathrm{d}t'\, \exp(-\mathrm{i}f(k)(t-t'))a(t'),$$

and, on inserting this into (4), we obtain the following equation of motion for $a(t)$.

$$\frac{\mathrm{d}a(t)}{\mathrm{d}t} = -\mathrm{i}\omega a(t) - \int_0^t \mathrm{d}t'\, \phi(t-t')a(t') + \xi(t), \tag{6}$$

where

$$\phi(t) = \int \mathrm{d}k\, |g(k)|^2 \exp(-\mathrm{i}f(k)t), \tag{7}$$

and

$$\xi(t) = \mathrm{i}\int \mathrm{d}k\, g(k)A(k)\exp(-\mathrm{i}f(k)t). \tag{8}$$

Since this last equation serves to express $\xi(t)$ in terms of the time-independent field operator $A(k)$, we see that the properties of $\xi(t)$ are determined, in statistical terms by the initial state Ψ of R – here it is only the initial state that is relevant, as we are describing the dynamics of $(S+R)$ in Heisenberg representation. Thus, $\xi(t)$ corresponds to a *stochastic* force and so (6) is a generalized, quantum mechanical *Langevin equation*† for the open system S. We remark here that $\xi(t)$ is not a Heisenberg operator for the whole system $(S+R)$, but only for R since, by (8), $\xi(t) = \mathrm{e}^{\mathrm{i}H_Rt}\xi(0)\mathrm{e}^{-\mathrm{i}H_Rt}$, where H_R is the reservoir Hamiltonian, given by the second term on the RHS of (3).

To formulate the statistical properties of $\xi(t)$, we invoke the vacuum property of the reservoir state Ψ (i.e. $A(k)\Psi = 0$) and the canonical commutation relations (2). Hence, by the definition (8) of $\xi(t)$, its

† Recall that the classical Langevin equation for the velocity $v(t)$ in a resistive medium takes the form $\mathrm{d}v/\mathrm{d}t = -\beta v + f(t)$ where β is a friction constant and $f(t)$ a fluctuating force, specified in statistical terms. In the generalized Langevin equation (6), the frictional force given by the second term on the RHS, is retarded.

expectation value and autocorrelation functions for the state Ψ are given by

$$\langle \xi(t) \rangle = 0 \tag{9}$$

$$\langle \xi(t)\xi^*(t') \rangle = \phi(t-t'); \quad \text{and} \quad \langle \xi^*(t)\xi(t') \rangle = \langle \xi(t)\xi(t') \rangle = 0, \tag{10}$$

where $\phi(t)$ is defined by (7). Thus, the retarded secular force on S, represented by the second term on the RHS of (6), and the statistical properties of the fluctuating force, $\xi(t)$, are both determined by the function $\phi(t)$. Furthermore, one sees from (7) that, in view of the Riemann–Lebesgue lemma,‡ this function decays to zero, as $t \to \infty$, for suitable forms of $f(k)$ and $g(k)$. Such decay corresponds to an irreversible loss of correlation of the 'force' $\xi^*(t)$ with $\xi(0)$ as time progresses, and leads to a corresponding irreversibility in the motion of S. To see how this arises in a simple way, we consider the case where

$$f(k) = k \quad \text{and} \quad g(k) = g_0, \quad \text{a constant.} \tag{11}$$

In this case, it follows from (7) that

$$\phi(t) = \kappa\delta(t), \quad \text{with} \quad \kappa = 2\pi g_0^2, \tag{12}$$

and consequently that the equation of motion (6) reduces to the standard Langevin form

$$\frac{\mathrm{d}a(t)}{\mathrm{d}t} = -(\mathrm{i}\omega + \kappa/2)a(t) + \xi(t) \tag{13}$$

where, by (9) and (10),

$$\langle \xi(t) \rangle = 0 \tag{14}$$

$$\langle \xi(t)\xi^*(t') \rangle = \kappa\delta(t-t'); \quad \text{and} \quad \langle \xi^*(t)\xi(t') \rangle = \langle \xi(t)\xi(t') \rangle = 0. \tag{15}$$

The R–S coupling is termed *singular* in cases where $\phi(t)$ has the δ-function form of (12). The simple feature of this coupling is that it carries no 'memory effects' in the Langevin operation for $a(t)$. We note that, by (7), singular coupling can arise only in cases where $f(k)$ runs through all real values from $-\infty$ to ∞. Physically, this means that modifications of Ψ can lead both to excitations and de-excitations of all energies, and thus that Ψ is not a ground state.

Assuming singular coupling, then, we deduce immediately from eqn (13) that

$$a(t) = a \exp -(\mathrm{i}\,\omega + \kappa/2)t + \int_0^t \mathrm{d}t' \exp(-(\mathrm{i}\omega + \kappa/2)(t-t'))\xi(t'), \tag{16}$$

and hence, denoting expectation values with respect to the initial state $\psi \otimes \Psi$ of $(R+S)$ by $\langle \ldots \rangle$, it follows from (14)–(16) that

$$\langle a(t) \rangle = \langle a \rangle \exp -(\mathrm{i}\omega + \kappa/2)t, \tag{17}$$

‡ This lemma states that, if $\phi(t) = \int_{-\infty}^{\infty} \mathrm{d}u \psi(u) \mathrm{e}^{\mathrm{i}ut}$, where $\int_{-\infty}^{\infty} \mathrm{d}u\, |\psi(u)|$ is finite, then $\phi(t) \to 0$ as $t \to \infty$ (cf. Whittaker and Watson 1946, pp. 172–4).

and

$$\langle a^*(t)a(t)\rangle = \langle a^*a\rangle \exp(-\kappa t). \tag{18}$$

These last two equations imply that the expectation values of the oscillator amplitude, $a(t)$, and its number of quanta, $a^*(t)a(t)$, relax irreversibly to zero as $t \to \infty$. Thus, R behaves as a sink, whose coupling to S damps down the oscillator motion.

Note. The singular coupling, though conveniently simple, is not essential for this result: what is needed is merely that $\phi(t)$ has suitable decay properties, that can be realized even when the frequency spectrum of the reservoir modes is non-negative.†

7.2.2. *Model of a pumped atom*

We formulate here a model of an atom, S, coupled to a reservoir, R, consisting of a pump and a sink. For simplicity, we take S to be a two-level atom, with single electron eigenstates θ_1 and θ_2, whose energies are ε_1 and $\varepsilon_2\,(>\varepsilon_1)$, respectively. We employ second quantization for the description of the atom, denoting the creation and annihilation operators for each state θ_j by b_j^* and b_j. These operators thus satisfy the canonical anticommutation relations

$$\{b_j^*, b_k\} = \delta_{jk}, \qquad \{b_j, b_k\} = 0, \tag{19}$$

where

$$\{A, B\} \equiv AB + BA,$$

and the Hamiltonian for S takes the form

$$H_S = \varepsilon_1 b_1^* b_1 + \varepsilon_2 b_2^* b_2. \tag{20}$$

The reservoir, R, is assumed to consist of four independent one-dimensional Fermi fields $B_{j,\pm}(k)$ $(j = 1, 2)$, with $B_{j,+}$ acting as a source, or pump, and $B_{j,-}$ as a sink for the electrons in the state θ_j. These fields are represented by operators satisfying the canonical anticommutation relations

$$\{B_{j,+}^*(k), B_{j,+}(k')\} = \{B_{j,-}^*(k), B_{j,-}(k')\} = \delta(k-k') \quad \text{for } j = 1, 2, \tag{21}$$

all other anticommutators between the B's and B^*'s, as well as those between these field operators and the electronic ones, b and b^*, vanishing. The Hilbert space $\mathcal{H}$ in which the fields B and B^* act is taken to be the Fock-type space, defined by the conditions that (a) $\mathcal{H}$ contains a vector Ψ such that

$$B_{j,+}^*(k)\Psi = B_{j,-}(k)\Psi = 0 \quad \text{for} \quad j = 1, 2, \tag{22}$$

which means that Ψ is full of B_+-quanta and void of B_--quanta; and (b) $\mathcal{H}$

† See Sewell (1974).

is generated by application to Ψ of all polynomials in $\int B_{j,+}(k)F_+(k)\,dk$ and $\int B^*_{j,-}(k)F_-(k)\,dk$, with $F_\pm(k)$ square integrable.

The reservoir Hamiltonian is assumed to be

$$H_R = \sum_{j=1,2} \int dk\, f(k)(-B_{j,+}(k)B^*_{j,+}(k) + B^*_{j,-}(k)B_{j,-}(k)), \tag{23}$$

where $f(k)$ is the frequency of the modes of wave vector k and the contributions to the integrand given by $-f(k)B_{j,+}(k)B^*_{j,+}(k)$ and $f(k)B^*_{j,-}(k)B_{j,-}(k)$ represent the energies of these modes of the B_+ and B_- fields, respectively, as measured relative to the state Ψ. The coupling between the atom, S, and the reservoir, R, is assumed to be given by the interaction Hamiltonian

$$H_{\text{int}} = \sum_{j=1,2} \int dk[(g_{j,+}(k)B^*_{j,+}(k) + g_{j,-}(k)B^*_{j,-}(k))b_j + \text{Hermitian conjugate}], \tag{24}$$

with $g_{j,\pm}(k)$ real-valued functions, representing the coupling of the atomic state θ_j to the fields $B_{j,\pm}$. The Hamiltonian for the composite system $(S+R)$ is then

$$H = H_S + H_R + H_{\text{int}} \tag{25}$$

Thus, by eqns (20) and (23)–(25), the model consists of Fermi oscillators, b_1 and b_2, harmonically coupled to Fermi fields, $B_{1,\pm}$ and $B_{2,\pm}$, respectively. It therefore has some obvious analogies with the oscillator model of §7.2.1, though in the present model the quanta are all fermions.

We assume that S is prepared in a pure state ψ and that R is in the state Ψ, characterized by the condition (22), when the two systems are coupled together at $t=0$. The initial state of the composite system $(S+R)$ is then $\psi \otimes \Psi$. Since, by (22), Ψ is full of B_+-quanta and void of B_--quanta, the fields $B_{j,\pm}$ act as source and sink, respectively, for the state θ_j.

We formulate the dynamics of the model in Heisenberg representation, denoting the operators corresponding to time translates of b_j and $B_{j,\pm}(k)$ by $b_j(t)$ $(\equiv e^{iHt}b_j e^{-iHt})$ and $B_{j,\pm}(k,t)$ $(\equiv e^{iHt}B_{j,\pm}(k)e^{-iHt})$, respectively. It follows from these definitions, together with the formula (25) for H and the canonical anticommutation relations (19) and (21), that

$$\frac{db_j(t)}{dt} = -i\varepsilon_j b_j(t) - i\int dk(g_{j,+}(k)B_{j,+}(k,t) + g_{j,-}(k)B_{j,-}(k,t)), \tag{26}$$

and

$$\frac{dB_{j,\pm}(k,t)}{dt} = -if(k)B_{j,\pm}(k,t) - ig_{j,\pm}(k)b_j(t). \tag{27}$$

On solving this last equation in the form

$$B_{j,\pm}(k,t)=B_{j,\pm}(k)\exp(-\mathrm{i}f(k)t)-\mathrm{i}g_{j,\pm}(k)\int_0^t \mathrm{d}t'\exp(-\mathrm{i}f(k)(t-t'))b_j(t'),$$

and substituting this formula into (26), we obtain the following equation of motion for $b_j(t)$.

$$\frac{\mathrm{d}b_j(t)}{\mathrm{d}t}=-\mathrm{i}\varepsilon_j b_j(t)-\int_0^t \mathrm{d}t'\ \phi_j(t-t')b_j(t')+\eta_j(t), \tag{28}$$

where

$$\phi_j(t)=\phi_{j,+}(t)+\phi_{j,-}(t) \tag{29}$$

$$\eta_j(t)=\eta_{j,+}(t)+\eta_{j,-}(t) \tag{30}$$

$$\phi_{j,\pm}(t)=\int \mathrm{d}k\ |g_{j,\pm}(k)|^2\exp(-\mathrm{i}f(k)t), \tag{31}$$

and

$$\eta_{j,\pm}(t)=-\mathrm{i}\int \mathrm{d}k\, g_{j,\pm}(k)B_{j,\pm}(k)\exp(-\mathrm{i}f(k)t). \tag{32}$$

Thus, (28) is a *generalized Langevin equation*, with $\eta_j(t)$ a stochastic force, whose statistical properties are determined by the initial state of R. Like its counterpart, $\xi(t)$, for the oscillator model of §7.2.1, this force is not a Heisenberg operator for the whole system $(S+R)$, but only for R, i.e. $\eta_j(t)=\mathrm{e}^{\mathrm{i}H_Rt}\eta_j(0)\mathrm{e}^{-\mathrm{i}H_Rt}$.

The statistical properties of $\eta_j(t)$ for the state Ψ are easily obtained from the characterization (22) of this state, together with the canonical anticommutation relations (19) and (21), in the following form.

$$\langle\eta_j(t)\rangle=0 \tag{33}$$

$$\langle\eta_j^*(t)\eta_k(t')\rangle=\phi_{j,+}(t-t')\delta_{jk};\quad \text{and}\quad \langle\eta_j(t)\eta_j^*(t')\rangle=\phi_{j,-}(t-t')\delta_{jk'}, \tag{34}$$

where the functions $\phi_{j,\pm}(t)$ are given by (31). Proceeding as in the treatment of the oscillator model of §7.2.1, we assume forms of $f(k)$ and $g_{j,\pm}(k)$ that lead to δ-function forms for $\phi_{j,\pm}(t)$ – again, this is the assumption of *singular coupling*. Thus, we assume that

$$f(k)=k\quad \text{and}\quad g_{j,\pm}(k)=g_{j,\pm},\quad \text{a constant}, \tag{35}$$

which implies, by (31), that

$$\phi_{j,\pm}(t)=\gamma_{j,\pm}\delta(t),\quad \text{with}\quad \gamma_{j,\pm}=2\pi\,|g_{j,\pm}|^2. \tag{36}$$

Hence, the generalized Langevin equation (28) reduces to the standard form

$$\frac{db_j(t)}{dt} = -(i\varepsilon_j + \tfrac{1}{2}\gamma_j)b_j(t) + \eta_j(t), \tag{37}$$

where

$$\gamma_j = \gamma_{j,+} + \gamma_{j,-}, \tag{38}$$

while the formulae (33) and (34) reduce to

$$\langle \eta_j(t)\rangle = 0;\ \langle \eta_j^*(t)\eta_k(t')\rangle = \gamma_{j,+}\delta(t-t')\delta_{jk};$$

$$\langle \eta_j(t)\eta_k^*(t')\rangle = \gamma_{j,-}\delta(t-t')\delta_{jk}. \tag{39}$$

The solution of the Langevin equation (37) is

$$b_j(t) = b_j \exp -(i\varepsilon_j + \gamma_j/2)t + \int_0^t dt' \exp(-(i\varepsilon_j + \gamma_j/2)(t-t'))\eta_j(t'). \tag{40}$$

Hence, denoting averages with respect to the initial state $\psi \otimes \Psi$ of $(S+R)$ by $\langle \ldots \rangle$, we see from (39) and (40) that

$$\langle b_j(t)\rangle = \langle b_j\rangle \exp -(i\varepsilon_j + \gamma_j/2)t \tag{41}$$

and

$$\langle b_j^*(t)b_j(t)\rangle = \langle b_j^* b_j\rangle e^{-\gamma_j t} + \left(\frac{\gamma_{j,+}}{\gamma_j}\right)(1 - e^{-\gamma_j t}), \tag{42}$$

which implies that the expectation values of $b_j(t)$ and $b_j^*(t)b_j(t)$ relax irreversibly to 0 and $\gamma_{j,+}/\gamma_j$, respectively. In particular, the terminal occupation probability of the state θ_j $(j = 1, 2)$ is

$$p_j = \frac{\gamma_{j,+}}{\gamma_j} \equiv \frac{\gamma_{j,+}}{\gamma_{j,+} + \gamma_{j,-}}, \tag{43}$$

and so is determined by the ratio of the parameters $\gamma_{j,\pm}$ representing the strengths of the couplings of this state to the fields $B_{j,\pm}$ of its pump and sink.

Dynamics of atomic observables. Although the solution (40) of the Langevin equation, together with the formula (39) for the stochastic properties of the fluctuating forces $\eta_j(t)$, completely determine the dynamics of the atom S, it is of interest, especially in connection with the laser model of the following Section, to formulate explicit equations of motion for the observables of this system given by

$$\nu = b_1^* b_1 + b_2^* b_2, \tag{44}$$

which represents the total number of electrons,

$$\rho = b_2^* b_2 - b_1^* b_1, \tag{45}$$

the difference between the number of them in the states θ_2 and θ_1, and

$$\sigma = b_1^* b_2 \quad \text{and} \quad \sigma^* = b_2^* b_1, \tag{46}$$

which represent the atomic polarization, subject to the assumption that the electrostatic potential for a uniform electric field has zero expectation values in the states θ_1 and θ_2. For simplicity, we shall restrict our formulation of the dynamics of these variables to situations where the couplings between atom and reservoir are adjusted so that the damping constants, γ_1 and γ_2, for the states θ_1 and θ_2 are equal, and the terminal expectation value of the number of electrons in the atom is unity, i.e.

$$\gamma_1 = \gamma_2 = \gamma, \text{ say}, \tag{47}$$

and

$$p_1 + p_2 = 1; \tag{48}$$

or equivalently, by (43),

$$\gamma_{2,+} = \gamma_{1,-} = \gamma(1 + \tfrac{1}{2}\tau) \quad \text{and} \quad \gamma_{2,-} = \gamma_{1,+} = \gamma(1 - \tfrac{1}{2}\tau), \tag{49}$$

where the parameter τ lies between -1 and 1. One sees from (43) and (49) that this parameter represents the difference $(p_2 - p_1)$ between the terminal occupation probabilities for the eigenstates θ_2 and θ_1 of the atom. Therefore, when positive, it represents the degree of *population inversion* of the atom, i.e. the extent to which the excited state is more populated than the ground one. It is usually referred to as the *pump parameter* of the model.

It is now a simple matter to derive the equations of motion for the Heisenberg operators $\nu(t)$, $\rho(t)$ and $\sigma(t)$, representing the time-translates of the atomic observables ν, ρ, and σ, defined by the formulae (44)–(47). For, on expressing these Heisenberg operators in terms of $b_j(t)$ and $b_j^*(t)$, it follows easily from these equations and (37) that their time-derivatives are given by the following formulae.

$$\frac{\mathrm{d}\nu(t)}{\mathrm{d}t} = -\gamma(\nu(t) - 1) + \chi_\nu(t), \tag{50}$$

$$\frac{\mathrm{d}\rho(t)}{\mathrm{d}t} = -\gamma(\rho(t) - \tau) + \chi_\rho(t), \tag{51}$$

and

$$\frac{\mathrm{d}\sigma(t)}{\mathrm{d}t} = -(\mathrm{i}\varepsilon + \gamma)\sigma(t) + \chi_\sigma(t), \quad \text{with} \quad \varepsilon = \varepsilon_2 - \varepsilon_1, \tag{52}$$

where

$$\chi_v(t) = \left(b_2^*(t)\eta_2(t) + b_1^*(t)\eta_1(t) - \frac{\gamma}{2}\right) + \text{Hermitian conjugate}, \tag{53}$$

$$\chi_\rho(t) = \left(b_2^*(t)\eta_2(t) - b_1^*(t)\eta_1(t) - \frac{\gamma\tau}{2}\right) + \text{Hermitian conjugate}, \tag{54}$$

and

$$\chi_\sigma(t) = b_2^*(t)\eta_1(t) + \eta_2^*(t)b_1(t), \tag{55}$$

Thus, although (50)–(52) simulate Langevin equations, the stochastic forces $\chi(t)$, that appear there, depend not only on the reservoir fields, $\eta(t)$ and $\eta^*(t)$, but also on the electronic operators, $b(t)$ and $b^*(t)$. Nevertheless, as we shall show in the Appendix, it follows from the canonical anticommutation relations, together with the characterization (22) of the reservoir state Ψ and the formula (40) for $b(t)$, that these forces $\chi(t)$ possess the following simple statistical properties.

$$\langle\chi_v(t)\rangle = \langle\chi_\rho(t)\rangle = \langle\chi_\sigma(t)\rangle = 0 \tag{56}$$

$$\langle v\chi_v(t)\rangle = \langle\rho\chi_\rho(t)\rangle = \langle\sigma^*\chi_\sigma(t)\rangle = 0 \tag{57}$$

$$\langle\chi_v(t)\chi_v(t')\rangle = \gamma(1 - \tau\bar{\rho}(t))\delta(t-t') \tag{58}$$

$$\langle\chi_\rho(t)\chi_\rho(t')\rangle = \gamma(1 - \tau\bar{\rho}(t))\delta(t-t') \tag{59}$$

and

$$\langle\chi_\sigma^*(t)\chi_\sigma(t')\rangle = \gamma(1 - \tfrac{1}{2}\tau)\bar{\rho}(t)\delta(t-t'), \tag{60}$$

where $\bar{\rho}(t)$ is the expectation value of $\rho(t)$ for the initial state.

It is now a simple matter to obtain the evolution of the expectation values and dispersions of the atomic observables from the equations of motion (50)–(52), together with the properties (56)–(60) of the stochastic forces χ. For, defining

$$\bar{v}(t) = \langle v(t)\rangle, \qquad \bar{\rho}(t) = \langle\rho(t)\rangle \quad \text{and} \quad \bar{\sigma}(t) = \langle\sigma(t)\rangle, \tag{61}$$

it follows from (50)–(52) that

$$\frac{\mathrm{d}\bar{v}(t)}{\mathrm{d}t} = -\gamma(v(t) - 1); \quad \frac{\mathrm{d}\bar{\rho}(t)}{\mathrm{d}t} = -\gamma(\rho(t) - \tau) \text{ and } \frac{\mathrm{d}\bar{\sigma}(t)}{\mathrm{d}t} = -(\gamma + \mathrm{i}\varepsilon)\bar{\sigma}(t), \tag{62}$$

which implies that $\bar{v}(t)$, $\bar{\rho}(t)$, and $\bar{\sigma}(t)$ relax to their terminal values, 1, τ, and 0, respectively, as $t \to \infty$. To obtain the dispersion of say, $v(t)$, we note that, by (50) and (62),

$$\frac{\mathrm{d}}{\mathrm{d}t}(v(t) - \bar{v}(t)) = -\gamma(v(t) - \bar{v}(t)) + \chi_v(t)$$

and therefore

$$v(t)-\bar{v}(t)=(v-\bar{v}(0))\mathrm{e}^{-\gamma t}+\int_0^t \mathrm{d}t'\mathrm{e}^{-\gamma(t-t')}\chi_v(t').$$

Hence, by (57) and (58), the time-dependent dispersion $\Delta v(t)$ of $v(t)$ is given by

$$(\Delta v(t))^2=\langle(v(t)-\bar{v}(t))^2\rangle=(\Delta v(0))^2\mathrm{e}^{-2\gamma t}+\gamma^2\int_0^t \mathrm{d}t'\mathrm{e}^{-2\gamma(t-t')}(1-\tau\bar{\rho}(t')). \tag{63}$$

It is a straightforward matter to obtain an explicit formula for $\Delta v(t)$ from this equation and (62) and thence to infer that this dispersion converges to a definite limit, that is independent of the initial state of S, as $t\to\infty$. Similarly, one may obtain explicit formulae for the time-dependent dispersions of $\rho(t)$ and $\sigma(t)$ and show that these, too, relax to terminal values that do not depend on the initial state of S.

7.3. The laser model†

A laser is a device that converts a random supply of energy, by pumping, into a monochromatic light beam, that is coherent in Glauber's (1963*b*) sense, i.e. the statistical properties of its electric field, $E(x,t)$, satisfy the condition that

$$\langle E(x_1,t_1)\ldots E(x_n,t_n)\rangle=\prod_1^n\langle E(x_j,t_j)\rangle, \tag{64}$$

where

$$\langle E(x,t)\rangle=E_0\cos(k\,.\,x-\omega t+\delta) \tag{65}$$

and E_0, ω, k, and δ are constants, the amplitude E_0 being macroscopic. Thus, a laser beam corresponds to a macroscopically excited radiation mode and so simulates the condensed phase of a Bose gas, where one has a macroscopic occupation of a single-particle state. By contrast, a normal light beam, even a seemingly monochromatic one, does not satisfy the condition (64), since it consists of an enormous number of components of different frequencies, whose interference destroys the correlations of the electric field at different space–time points.

The laser generally consists of a system of atoms interacting with an electromagnetic field, of which they are the source, in a cavity, together with reservoirs for both the atoms and the field. The atomic reservoirs consist of pumps and sinks: the field reservoirs are simply sinks. The atomic pumps consist of irradiation, that serves to excite the atoms. The sinks of both the atom and the field are provided by the rest of the apparatus, e.g. the cavity walls, that serve to absorb their energy.

† See Haken's (1970) book for an authoritative account of the basic principles of laser theory.

Remarkably, the random supply of energy to the atoms is converted into coherent radiation when the pumping rate exceeds a certain critical value. In fact, one has a phase transition from normal to coherent light when the pumping rate passes beyond this threshold value. The mechanism by which this transition occurs can be understood on the basis of a relatively simple model, due originally to Dicke (1954). We shall now describe the theory of the phase transition of this model, as given by the rigorous treatment of Hepp and Lieb (HL) (1973).

The model. This consists of N atoms $A_1, \ldots, A_N$, interacting with an electromagnetic field, F, in a cavity of volume V, together with reservoirs for the atoms and field. The components of this model, which we denote by Σ_N, and their interactions may be described as follows.

(a) The atoms, $A_1, \ldots, A_N$ are identical, except for their locations in the cavity. It is assumed that each of these, together with its reservoir, is a copy of that of §7.2.2, with singular coupling, and that its interaction parameters satisfy the conditions (49), designed to produce a population of one electron per atom. We denote by $b_{l,j}$ and $\eta_{l,j,\pm}(t)$ the counterparts, for A_l, of the electron annihilation operator b_j and the random force $\eta_{j,\pm}(t)$ of the model of §7.2.2; and we assume that these operators anticommute with all the corresponding fermion operators and their adjoints for any other atom, A_m. We denote by v_l, ρ_l and σ_l the operators representing the total number of electrons in A_l, the difference between the numbers of these in its excited and ground states and the atomic polarization, i.e. (cf. eqns (44)–(46))

$$v_l = \sum_{j=1,2} b_{l,j}^* b_{l,j} \tag{66}$$

$$\rho_l = \sum_{j=1,2} (-1)^j b_{l,j}^* b_{l,j} \tag{67}$$

$$\sigma_l = b_{l,1}^* b_{l,2} \quad \text{and} \quad \sigma_l^* = b_{l,2}^* b_{l,1}. \tag{68}$$

(b) The field, F, is represented as a *single mode*, or oscillator. It is assumed that this mode, together with its reservoir, $R(F)$, is a copy of the oscillator model of §7.2.1, with singular coupling, and that its frequency, ω, corresponds to the difference between the energies, ε_2 and ε_1, of the states of each atom A_l, i.e.

$$\omega = \varepsilon \equiv \varepsilon_2 - \varepsilon_1, \tag{69}$$

a resonance condition. Using the same notation as in §7.2.1, we denote the creation and annihilation operators for the oscillator, F, by a^* and a, respectively, and the random force on it, due to the reservoir, by $\xi(t)$. We assume that these operators commute with all those of the atoms, A_l, and their reservoirs, $R(A_l)$.

(c) The coupling between the atoms and the field, F, is dipolar and is given by the following interaction Hamiltonian:

$$H^{FA} = \lambda N^{-\frac{1}{2}} \sum_{l=1}^{N} (a^* b_{l,1}^* b_{l,2} + a b_{l,2}^* b_{l,1}) \equiv \lambda\, N^{-\frac{1}{2}} \sum_{1}^{N} (a^* \sigma_l + a \sigma_l^*), \tag{70}$$

where λ is a coupling constant and the factor $N^{-\frac{1}{2}}$ ($\sim V^{-\frac{1}{2}}$) arises from the decomposition of the electromagnetic field into normal modes before all but one of these is discarded.

(d) The field F, the atoms $A_1, \ldots, A_N$ and the reservoirs $R(F)$, $R(A_1), \ldots, R(A_N)$ are prepared, independently of one another, in states ψ_{F}, $\psi_{A_1}, \ldots, \psi_{A_N}$, Ψ_F, $\Psi_{A_1}, \ldots, \Psi_{A_N}$, respectively, where Ψ_F and Ψ_{A_l} $(l = 1, \ldots, N)$ are copies of the initial states of the reservoirs attached to the oscillator and atom of §§7.2.1 and 7.2.2 respectively. Thus, the initial state of the entire system Σ_N is

$$\Phi^{(N)} = \psi_F \otimes \psi_A \otimes \Psi_F \otimes \Psi_A, \tag{71}$$

where

$$\psi_A = \bigotimes_{1}^{N} \psi_{A_l} \quad \text{and} \quad \Psi_A = \bigotimes_{1}^{N} \Psi_{A_l}. \tag{72}$$

Macroscopic description. Assuming N to be a 'large' extensivity parameter, we formulate the macroscopic description of the model in terms of the intensive variables

$$\nu^{(N)} = N^{-1} \sum_{1}^{N} \nu_l \tag{73}$$

$$\rho^{(N)} = N^{-1} \sum_{1}^{N} \rho_l \tag{74}$$

and

$$\sigma^{(N)} = N^{-1} \sum_{1}^{N} \sigma_l, \tag{75}$$

representing the densities of atomic number, population inversion and polarization, respectively, together with the rescaled field operator

$$\alpha^{(N)} = N^{-\frac{1}{2}} a, \tag{76}$$

the factor $N^{-\frac{1}{2}}$ being designed to render $\alpha^{(N)}$ intensive in situations where the field energy, $\omega a^* a$, has the magnitude of an extensive quantity. We note that it follows from the definitions of $\sigma^{(N)}$ and $\alpha^{(N)}$ given by the last two equations, that the formula (70) for H_{FA} is equivalent to the following:

$$H^{FA} = N\lambda(\alpha^{(N)*}\sigma^{(N)} + \sigma^{(N)*}\alpha^{(N)}). \tag{70$'$}$$

Our basic objective is to extract the dynamics of the macroscopic variables $\nu^{(N)}$, $\rho^{(N)}$, $\sigma^{(N)}$, and $\alpha^{(N)}$ from the microscopic equations of the motion of the entire system, Σ_N, in the limit $N \to \infty$. This effectively means that we are dealing with a sequence $\{\Sigma_N\}$ of versions of the model described here.

Let us begin with some general observations concerning the classical character of the macroscopic observables. Thus, we first note that, by eqns (73)–(76) and our above specifications (a) and (b) of the model,

$$[\rho^{(N)}, \sigma^{(N)}] = -2N^{-1}\sigma^{(N)}; \quad \text{and} \quad [\sigma^{(N)*}, \sigma^{(N)}] = N^{-1}\rho^{(N)}, \tag{77}$$

all other commutators between the macrovariables vanishing. Hence, as the norms of $\sigma^{(N)}$ and $\rho^{(N)}$ are uniformly bounded, by (67), (68), (74), and (75), it follows that the observables $\sigma^{(N)}$, $\rho^{(N)}$, $\nu^{(N)}$, and $\alpha^{(N)}$ intercommute in the limit $N \to \infty$. They therefore simulate *classical* variables in that limit.

Next, we observe that, as the initial atomic state $\psi_A = \otimes_1^N \psi_{A_l}$ carries no correlations between the observables of different atoms, it follows from the definitions (73)–(75) of $\nu^{(N)}$, $\rho^{(N)}$, and $\sigma^{(N)}$ that their initial dispersions are $0(N^{-\frac{1}{2}})$. By (76), the same will be true for the field variable $\alpha^{(N)}$ if, as we shall assume, the dispersion of the field operator a, in the initial state ψ_F, is $0(1)$. Thus, denoting the dispersion of an arbitrary observable X by ΔX,

$$\Delta\alpha^{(N)}, \Delta\nu^{(N)}, \Delta\rho^{(N)} \text{ and } \Delta\sigma^{(N)} \to 0 \quad \text{as} \quad N \to \infty, \text{ at } t = 0. \tag{78}$$

We also assume that the initial expectation values of $\alpha^{(N)}$, $\nu^{(N)}$, $\rho^{(N)}$ and $\sigma^{(N)}$ converge to definite limits as $N \to \infty$, i.e.

$$\langle \alpha^{(N)} \rangle, \langle \nu^{(N)} \rangle, \langle \rho^{(N)} \rangle, \langle \sigma^{(N)} \rangle \to \alpha, \nu, \rho, \sigma, \text{ respectively,}$$
$$\text{as } N \to \infty. \tag{78$'$}$$

The last two equations signify that the macrovariables $\alpha^{(N)}$, $\nu^{(N)}$, $\rho^{(N)}$, and $\sigma^{(N)}$ are initially *sharply defined*, with values α, ν, ρ, and σ, respectively, in the limit $N \to \infty$. As we shall presently show, these variables subsequently evolve, in this limit, according to a self-contained classical, irreversible, deterministic law, that exhibits a phase transition of the radiation mode, F, when the atomic pumping rate passes through a critical value.

Microscopic equations of motion. By the above specifications (a)–(d) of Σ_N, the Heisenberg equations of motion for the wave operator a and the atomic observables ν_l, ρ_l and σ_l take the forms given by eqns (13) and (50)–(52) for the oscillator and atomic models of §7.2, together with additional contributions due to the coupling H^{FA} between the field and the atoms. In fact, for any observable X, the contribution to $\mathrm{d}X(t)/\mathrm{d}t$, due to this coupling, is simply $\mathrm{i}[H^{FA}, X](t)$. Thus, by the formula (70) for H^{FA} and the canonical commutation and anticommutation relations for a, a^* and b, b^*, respectively, we obtain the following equations of motion.

$$\frac{\mathrm{d}a(t)}{\mathrm{d}t} = -(\mathrm{i}\varepsilon + \kappa/2)a(t) - \mathrm{i}\lambda N^{-\frac{1}{2}} \sum_1^N \sigma_l(t) + \xi(t), \tag{79}$$

the oscillator frequency being ε, by (69),

$$\frac{\mathrm{d}\nu_l(t)}{\mathrm{d}t} = -\gamma(\nu_l(t)-1)+\chi_{\nu,l}(t) \tag{80}$$

$$\frac{\mathrm{d}\rho_l(t)}{\mathrm{d}t} = -\gamma(\rho_l(t)-\tau)+2\mathrm{i}\lambda N^{-\frac{1}{2}}(a^*(t)\sigma_l(t)-a_l(t)\sigma_l^*(t))+\chi_{\rho,l}(t) \tag{81}$$

and

$$\frac{\mathrm{d}\sigma_l(t)}{\mathrm{d}t} = -(\gamma+\mathrm{i}\varepsilon)\sigma_l(t)-\mathrm{i}\lambda N^{-\frac{1}{2}}a_l(t)\rho_l(t)+\chi_{\sigma,l}(t), \tag{82}$$

where the stochastic properties of $\xi(t)$ are given by (14) and (15), i.e.

$$\langle\xi(t)\rangle=\langle\xi^*(t)\xi(t')\rangle=0;\quad \text{and}\quad \langle\xi(t)\xi^*(t')\rangle=\kappa\delta(t-t'); \tag{83}$$

while $\chi_{\nu,l}$, $\chi_{\rho,l}$ and $\chi_{\sigma,l}$ are the stochastic forces obtained by replacing b_j, η_j by $b_{l,j}$, $\eta_{l,j}$, respectively, in the formulae (53)–(55) for the corresponding forces of the atomic model of §7.2.2. As we shall prove in the Appendix, these forces in Σ_N possess the properties (56) and (58)–(60) of their counterparts for the atomic system and furthermore, the χ's at different sites are uncorrelated despite the coupling between the atoms and the field mode.
Thus,

$$\langle\chi_{\nu,l}(t)\rangle=\langle\chi_{\rho,l}(t)\rangle=\langle\chi_{\sigma,l}(t)\rangle=0, \tag{84}$$

$$\langle\chi_{\nu,l}(t)\chi_{\nu,m}(t')\rangle=\gamma(1-\tau\bar{\rho}_l(t)\delta_{lm}\delta(t-t'), \tag{85}$$

$$\langle\chi_{\rho,l}(t)\chi_{\rho,m}(t')\rangle=\gamma(1-\tau\bar{\rho}_l(t))\delta_{lm}\delta(t-t'), \tag{86}$$

and

$$\langle\chi_{\sigma,l}^*(t)\chi_{\sigma,m}(t')\rangle=\gamma(1-\tfrac{1}{2}\tau)\bar{\rho}_l(t)\delta_{lm}\delta(t-t'), \tag{87}$$

where $\bar{\rho}_l(t)$ is the expectation value of $\rho_l(t)$.

Macroscopic equations of motion. It now follows from eqns (79)–(82), together with the definitions (73)–(76) of the macroscopic variables $\alpha^{(N)}$, $\nu^{(N)}$, $\rho^{(N)}$, and $\sigma^{(N)}$, that the Heisenberg equations of motion of these are the following.

$$\frac{\mathrm{d}\alpha^{(N)}(t)}{\mathrm{d}t} = -(\mathrm{i}\varepsilon+\kappa/2)\alpha^{(N)}(t)-\mathrm{i}\lambda\sigma^{(N)}(t)+\xi^{(N)}(t), \tag{88}$$

$$\frac{\mathrm{d}\nu^{(N)}(t)}{\mathrm{d}t} = -\gamma(\nu^{(N)}(t)-1)+\chi_\nu^{(N)}(t), \tag{89}$$

$$\frac{\mathrm{d}\rho^{(N)}(t)}{\mathrm{d}t} = -\gamma(\rho^{(N)}(t)-\tau)+2\mathrm{i}\lambda(\alpha^{(N)*}(t)\sigma^{(N)}(t) - \alpha^{(N)}(t)\sigma^{(N)*}(t))+\chi_\rho^{(N)}(t), \tag{90}$$

and

$$\frac{\mathrm{d}\sigma^{(N)}(t)}{\mathrm{d}t} = -(\gamma + \mathrm{i}\varepsilon)\sigma^{(N)}(t) + \mathrm{i}\lambda\alpha^{(N)}(t)\rho^{(N)}(t) + \chi_\sigma^{(N)}(t), \tag{91}$$

where

$$\xi^{(N)}(t) = N^{-\frac{1}{2}}\xi(t) \quad \text{and} \quad \chi_\mu^{(N)}(t) = N^{-1}\sum_1^N \chi_{\mu,l}(t) \text{ for } \mu = \nu, \rho, \sigma. \tag{92}$$

From this last formula and eqns (74) and (83)–(87), we see that the fluctuating forces $\xi^{(N)}(t)$ and $\chi^{(N)}(t)$ have the following stochastic properties.

$$\langle \xi^{(N)}(t) \rangle = \langle \chi_\nu^{(N)}(t) \rangle = \langle \chi_\rho^{(N)}(t) \rangle = \langle \chi_\sigma^{(N)}(t) \rangle = 0, \tag{93}$$

$$\langle \xi^{(N)*}(t)\xi^{(N)}(t') \rangle = 0; \qquad \langle \xi^{(N)}(t)\xi^{(N)*}(t') \rangle = N^{-1}\kappa\delta(t - t'), \tag{94}$$

$$\langle \chi_\nu^{(N)}(t)\chi_\nu^{(N)}(t') \rangle = N^{-1}\gamma(1 - \tau\bar{\rho}^{(N)}(t))\delta(t - t'), \tag{95}$$

$$\langle \chi_\rho^{(N)}(t)\chi_\rho^{(N)}(t') \rangle = N^{-1}\gamma(1 - \tau\bar{\rho}^{(N)}(t))\delta(t - t'), \tag{96}$$

and

$$\langle \chi_\sigma^{(N)*}(t)\chi_\sigma^{(N)}(t') \rangle = N^{-1}\gamma(1 - \tfrac{1}{2}\tau)\bar{\rho}^{(N)}(t)\delta(t - t'), \tag{97}$$

where $\bar{\rho}^{(N)}(t)$ is the expectation value of $\rho^{(N)}(t)$. Hence, the means and autocorrelations of the stochastic forces $\xi^{(N)}(t)$ and $\chi^{(N)}(t)$ all vanish in the limit $N \to \infty$. Therefore since, in this limit, the macroscopic variables $\alpha^{(N)}(t)$, $\nu^{(N)}(t)$, $\rho^{(N)}(t)$ and $\sigma^{(N)}(t)$ intercommute and have sharply defined values for the state in which Σ_N is prepared, one can prove† that they reduce there to *classical variables* α_t, ν_t, ρ_t, and σ_t, whose equations of motion are obtained by discarding the forces $\xi^{(N)}(t)$ and $\chi^{(N)}(t)$ in (88)–(91), i.e.

$$\frac{\mathrm{d}\alpha_t}{\mathrm{d}t} = -(\kappa/2 + \mathrm{i}\varepsilon)\alpha_t - \mathrm{i}\lambda\sigma_t, \tag{98}$$

$$\frac{\mathrm{d}\nu_t}{\mathrm{d}t} = -\gamma(\nu_t - 1), \tag{99}$$

$$\frac{\mathrm{d}\rho_t}{\mathrm{d}t} = -\gamma(\rho_t - \tau) + 2\mathrm{i}\lambda(\alpha_t^*\sigma_t - \alpha_t\sigma_t^*), \tag{100}$$

and

$$\frac{\mathrm{d}\sigma_t}{\mathrm{d}t} = -(\gamma + \mathrm{i}\varepsilon)\sigma_t + \mathrm{i}\lambda\rho_t\alpha_t. \tag{101}$$

Specifically, the expectation values of the time-dependent macroscopic

† See Hepp and Lieb (1973).

observables reduce to their classical counterparts and their dispersions vanish in the limit $N \to \infty$. Similarly, expectation values of products of macroscopic variables at different times reduce to the products of the corresponding classical variables, e.g.

$$\lim_{N\to\infty} \langle \alpha^{(N)\#}(t_1) \ldots \alpha^{(N)\#}(t_n) \rangle = \alpha^{\#}_{t_1} \ldots \alpha^{\#}_{t_n}, \tag{102}$$

where each $\alpha^{\#}$ is either α or α^*. On comparing this formula with eqn (64), we see immediately that any solution of the classical eqns (98)–(101) for which α_t has a non-trivial sinusoidal form corresponds to coherent radiation. The problem then is to obtain conditions under which these equations have dynamically stable solutions of this kind.

The phase transition. As we have just seen, the macroscopic dynamics of the model is given by the classical equations of motion (98)–(101). We shall now show that these have a stable solution that changes from a non-radiating phase ($\alpha_t = 0$) to a coherently radiating one (α_t sinusoidal) when the pump parameter τ passes through a certain critical value.

Thus, we first note that, by (99), the number density, ν_t, evolves independently of the other macrovariables and relaxes exponentially to its terminal value of unity. This variable can therefore be dismissed from further consideration. We are left, then, with a classical dynamical system represented by the variables α_t, ρ_t, and σ_t that evolve according to eqns (98), (100), and (101). One sees easily that these have a unique stationary solution, namely

$$\alpha_t^{(0)} = 0, \qquad \rho_t^{(0)} = \tau, \qquad \sigma_t^{(0)} = 0. \tag{103}$$

To investigate its stability, we derive the linearized equations of motion for the deviations of α_t, ρ_t, and σ_t from this solution. Thus, defining

$$\delta\alpha_t = \alpha_t, \qquad \delta\rho_t = \rho_t - \tau \quad \text{and} \quad \delta\sigma_t = \sigma_t, \tag{104}$$

we deduce from (98), (100), and (101) that the required linear equations are

$$\frac{\mathrm{d}}{\mathrm{d}t}\delta\alpha_t = -\left(\frac{\kappa}{2} + \mathrm{i}\varepsilon\right)\delta\alpha_t - \mathrm{i}\lambda\gamma\sigma_t, \tag{105}$$

$$\frac{\mathrm{d}}{\mathrm{d}t}\delta\rho_t = -\gamma\delta\rho_t, \tag{106}$$

and

$$\frac{\mathrm{d}}{\mathrm{d}t}\delta\sigma_t = -(\gamma + \mathrm{i}\varepsilon)\delta\sigma_t + \mathrm{i}\lambda\tau\delta\alpha_t. \tag{107}$$

The second of these equations tells us that $\delta\rho_t$ relaxes exponentially to zero, and so stability for the variable ρ is guaranteed. To test for stability w.r.t. α

and σ, it suffices to look at the normal solutions of (105) and (107) for which $\delta\alpha_t$ and $\delta\sigma_t$ vary as e^{-zt} for some constant z. Thus, equating the time-derivatives of these functions to $-z\delta\alpha_1$ and $-z\delta\sigma_t$, respectively, we obtain the secular equation

$$(z+\kappa/2+i\varepsilon)(z+\gamma+i\varepsilon)-\lambda^2\tau=0,$$

whose roots are $z=-i\varepsilon+\frac{1}{2}\{(\gamma+\kappa/2)\pm((\gamma-\kappa/2)^2+4\lambda^2\tau)^{\frac{1}{2}}\}$. The condition for stability of the stationary solution (103) of the macroscopic equations is simply that the real parts of these roots are both positive. Hence, it is stable if and only if

$$\tau<\tau_c\equiv\kappa\gamma/2\lambda^2. \tag{108}$$

Thus, as τ increases through τ_c, the stationary solution loses its stability. However, the classical equations (98), (100), and (101) then acquire a periodic solution, namely

$$\left.\begin{aligned} &\alpha_t^{(1)}=Ae^{-i\varepsilon t}, \qquad \rho_t^{(1)}=\tau_c \quad \text{and} \quad \sigma_t^{(1)}=\frac{i\kappa A}{2\lambda}e^{-i\varepsilon t},\\ &\text{where}\\ &|A|=\left(\frac{\gamma}{2\kappa}(\tau-\tau_c)\right)^{\frac{1}{2}}, \end{aligned}\right\} \tag{109}$$

and the phase of A is undetermined. Moreover, this periodic solution is stable, as can be shown by linearizing the equations of motion for α_t, ρ_t, and σ_t about it. The stability analysis becomes particularly transparent in the case where $\kappa=\gamma$. For then, defining

$$\beta_t=\alpha_t e^{i\varepsilon t}, \tag{110}$$

it follows easily from eqns (98) and (100) that

$$\left(\frac{d}{dt}+\gamma\right)(2|\beta_t|^2+\rho_t)=\gamma\tau, \tag{111}$$

and from (98) and (101) that

$$\frac{d^2\beta_t}{dt^2}+\frac{3}{2}\gamma\frac{d\beta_t}{dt}+\frac{\gamma^2}{2}\beta_t-\lambda^2\rho_t\beta_t=0. \tag{112}$$

Since, by (111), $2|\beta_t|^2+\rho_t$ relaxes exponentially to τ we may replace ρ_t by $\tau-2|\beta_t|^2$ in (111). Hence, taking account of the assumption that $\kappa=\gamma$ and thus, by (108), that $\tau_c=\gamma^2/2\lambda^2$, we arrive at the equation of motion

$$\frac{d^2\beta_t}{dt^2}+\frac{3}{2}\gamma\frac{d\beta_t}{dt}+\frac{\partial}{\partial\beta_t}V(\beta_t,\beta_t^*), \tag{113}$$

where

$$V(\beta,\beta^*)\equiv\lambda^2((\beta\beta^*)^2-(\tau-\tau_c)\beta\beta^*)\equiv\lambda^2(|\beta|^4-(\tau-\tau_c)|\beta|^2). \tag{114}$$

Equation (113) represents a damped motion in a potential V, and the stable equilibrium points therefore correspond to the minima of V. By (114), these are given by $\beta = 0$ if $\tau < \tau_c$ and $|\beta| = (\frac{1}{2}(\tau - \tau_c))^{\frac{1}{2}}$ if $\tau > \tau_c$. Hence, in view of the relation (110) between β_t and α_t we see explicitly how, in the case $\gamma = \kappa$, the stable solution of the classical macroscopic equations changes from the stationary one, (103), to the periodic one, (109), as the pump parameter τ increases through the value τ_c.

We remark here that the transition from a stable stationary point to a stable periodic orbit, as a control parameter passes through a critical value, is a very general phenomenon, known as a *Hopf bifurcation*, in classical dissipative systems (cf. Hopf 1942).

For the present laser model, this transition involves a symmetry breakdown, since the equations of motion (98)–(101) are invariant under the transformations $\alpha \to \alpha e^{i\theta}$, $\sigma \to \sigma e^{i\theta}$, with θ constant, whereas the periodic orbit (109) evidently depends on the phase angle of α and σ. Thus, the transition simulates order–disorder transitions of equilibrium states. Furthermore, in the laser model, the field amplitude of A of the formula (109) plays the role here of an order parameter.

The coherent radiation. It follows immediately from the discussion following eqn (102) that the radiation in the ordered phase is coherent. We emphasize here that the coherence property is an essentially collective phenomenon arising from the couplings between the macroscopic variables $\alpha^{(N)}$, $\rho^{(N)}$, and $\sigma^{(N)}$ under appropriate pumping conditions. Its collective nature is evident from the fact that the classical deterministic character of the dynamics of these variables, which governs the coherence, depends crucially on the suppression of the stochastic forces $\xi^{(N)}(t)$, $\chi^{(N)}(t)$, and that is only achieved for enormously large (ideally infinite) N, by (94)–(97). Clearly, the coherence cannot be properly conceived in terms of successions of atomic emissions and absorptions of phonons.

7.4. Fröhlich's pumped phonon model†

This is a model of the conversion of randomly supplied energy into ordered structures in certain biological systems. It consists of N phonon modes, corresponding to polarization waves, that are supplied with energy, by pumping, and that exchange energy, both with one another and with a thermal reservoir, by emission and absorption of quanta. Remarkably, the combined effects of the pumping and the thermally activated exchanges of quanta leads to a *Bose–Einstein condensation of phonons* when the pumping is sufficiently strong. The polarization mode of lowest frequency is thus macroscopically excited. As we shall see, the basic reason why the

† See H. Fröhlich (1968, 1980, 1983) for discussion of the biological significance of the model.

pumped phonon system undergoes Bose–Einstein condensation, whereas it would merely take up a Planck distribution if coupled only to a heat bath, is that the competition between pumping and thermally activated exchanges of quanta essentially fixes the total number of phonons, so that the system simulates an ideal Bose gas.

The model. This consists of N phonon modes, whose frequencies are $\omega_1, \ldots, \omega_N$, and whose corresponding occupation numbers are denoted by $n_1, \ldots, n_N$, respectively. The model is formulated at the kinetic theoretic level and it is assumed that the rate of change of the occupation number, n_j, of the jth mode consists of three parts, namely, (a) a constant term s_j (>0), due to pumping, (b) terms proportional to $-n_j$ and $(1+n_j)e^{-\beta\omega_j}$, representing the emission and absorption of quanta, due to the coupling of the system to a thermal reservoir, and (c) terms proportional to $-n_j(1+n_k)e^{-\beta\omega_k}$ and $(1+n_j)n_k e^{-\beta\omega_j}$, due to the exchange of quanta with the kth mode, for all $k \neq j$. Thus, the terms (b) and (c) satisfy the condition of detailed balance.

It follows from these specifications that the kinetic equation of the model is given by

$$\frac{dn_j}{dt} = s_j - a_j(n_j - (1+n_j)e^{-\beta\omega_j})$$
$$- \sum_{k=1}^{N} b_{jk}(n_j(1+n_k)e^{-\beta\omega_j} - n_k(1+n_j)e^{-\beta\omega_k}), \qquad (115)$$

where a_j and b_{jk} are positive constants. For simplicity, we assume that $a_j(1-e^{-\beta\omega_j})$, b_{jk}, and $s_j + a_j e^{-\beta\omega_j}$ are independent of j and k, so that this kinetic equation, when expressed in appropriate units, reduces to the form

$$\frac{dn_j}{dt} = s - n_j - \frac{g}{N}\sum_{k=1}^{N} (n_j(1+n_k)e^{-\beta\omega_k} - n_k(1+n_j)e^{-\beta\omega_j}), \qquad (116)$$

where s and g are positive constants, and the factor N^{-1} in the last term is designed to make that intensive, when N becomes arbitrarily large. We define the intensive variable

$$\nu = N^{-1}\sum_{1}^{N} n_j, \qquad (117)$$

which represents the density of quanta, assuming N to be a measure of the volume of the system. It follows immediately from (116) and (117) that

$$\frac{d\nu}{dt} = s - \nu, \qquad (118)$$

which implies that ν relaxes exponentially to the value s. Hence, *the terminal density of phonons in the model is fixed by the pump parameter s.*

A second intensive variable that arises in the kinetics of the model is

$$\theta = N^{-1} \sum_{1}^{N} (n_j + 1)e^{-\beta\omega_j}. \tag{119}$$

For, by (117) and (119), the kinetic equation (116) may be expressed in the form

$$\frac{dn_j}{dt} = s - n_j - g\theta n_j + gv(1 + n_j)e^{-\beta\omega_j}. \tag{120}$$

The stationary distribution. For a stationary distribution, it follows from this last equation that

$$s - (1 + g\theta)n_j + gv(1 + n_j)e^{-\beta\omega_j} = 0, \tag{121}$$

and from (118) that $v = s$. Hence

$$n_j = \frac{s(1 + ge^{-\beta\omega_j})}{1 + g\theta - gse^{-\beta\omega_j}}, \tag{122}$$

where θ must take a value for which the relations (117) and (119) are satisfied. In fact, these relations, taken together with (122), provide equivalent conditions for θ, since, on summing equation (121) over j and putting $v = s$, it follows that

$$s - (1 + g\theta)N^{-1} \sum_{1}^{N} n_j + gsN^{-1} \sum_{1}^{N} (n_j + 1)e^{-\beta\omega_j} = 0,$$

and thus that if $N^{-1} \sum n_j = v\ (= s)$ then $N^{-1} \sum_1^N (1 + n_j)e^{-\beta\omega_j} = \theta$, and vice versa. Hence, a stationary distribution is given by (122), together with (117) with $v = s$ (or equivalently by (119)) as a self-consistence condition for θ. This distribution may conveniently be expressed in terms of the parameter

$$\mu = \beta^{-1}\ln(gs/1 + g\theta), \tag{123}$$

instead of θ. For, by this equation and (122),

$$n_j = \frac{1 + g^{-1}e^{\beta\omega_j}}{\exp \beta(\omega_j - \mu) - 1}, \tag{124}$$

and the self-consistency condition (117) for θ, i.e. for μ, with $v = s$, takes the form

$$s = N^{-1} \sum_{1}^{N} \frac{1 + g^{-1}e^{\beta\omega_j}}{\exp \beta(\omega_j - \mu) - 1}. \tag{125}$$

From eqn (124), we see that n_j is the stationary distribution for an ideal Bose gas, modified by the factor $(1 + g^{-1}e^{\beta\omega_j})$, with μ playing the role of a chemical potential; while (125) represents the condition that the total density of quanta is fixed at the value s. To show that equations (124) and

(125) uniquely define n_j and μ in terms of β and s, we first observe that the positivity of all the n's ensures that

$$\mu > \omega_1, \tag{126}$$

where ω_1 is the lowest phonon frequency. Moreover, the function of μ given by the RHS of the condition (125) decreases monotonically from ∞ to 0 as ν runs from ω_1 to ∞, and therefore that condition is satisfied by precisely one value of μ in the allowed range (ω_1, ∞). We conclude therefore that, for given s and β, there is a unique stationary distribution and that is defined by the formulae (124) and (125).

Stability. To show that this distribution is stable, we linearize the kinetic equation (120) around it: here, we may replace ν by s, as ν relaxes exponentially to this value. Thus, letting $n_j \to n_j + \delta n_j$ and $\theta \to \theta + \delta\theta$ in (120), with $\nu = s$, and linearizing, we obtain the equation

$$\frac{\mathrm{d}}{\mathrm{d}t}\delta n_j = -(1 + g\theta - gs\mathrm{e}^{-\beta\omega_j})\delta n_j - gn_j\delta\theta, \tag{127}$$

where, by (119),

$$\delta\theta = N^{-1}\sum_{1}^{N} \delta n_j \mathrm{e}^{-\beta\omega_j}, \tag{128}$$

and n_j, θ are the values of these variables for the stationary distribution. To test for stability, it suffices to consider the normal exponential solutions, $\delta n_j \sim \mathrm{e}^{-\lambda t}$, of (127). Thus, putting $(\mathrm{d}/\mathrm{d}t)\delta n_j = -\lambda\delta n_j$ there and using eqn (123) to express θ in terms of μ, we find that

$$\delta n_j = -gn_j\delta\theta/(gs\mathrm{e}^{-\beta\mu} - gs\mathrm{e}^{-\beta\omega_j} - \lambda);$$

and this formula, together with (128), implies that λ satisfies the secular equation

$$1 + N^{-1}\sum_{1}^{N} n_j\mathrm{e}^{-\beta\omega_j}/(s\mathrm{e}^{-\beta\mu} - s\mathrm{e}^{-\beta\omega_j} - g^{-1}\lambda) = 0. \tag{129}$$

Allowing for the possibility that λ might be complex, we note that the imaginary part of this equation is given by

$$(\mathrm{Im}\,\lambda)(gN)^{-1}\sum_{1}^{N} n_j\mathrm{e}^{-\beta\omega_j}/|s\mathrm{e}^{-\beta\mu} - s\mathrm{e}^{-\beta\omega_j} - g^{-1}\lambda|^2 = 0,$$

which implies that $\mathrm{Im}\,\lambda = 0$, i.e. that the roots of the secular equation are all real. Further, they must also be positive, since the condition (126) implies that, if λ were $\leqslant 0$, then every term on the RHS of (129) would be positive, and so that equation would not be satisfied. The positivity of the roots of the secular equation implies the stability of the stationary distribution.

Bose–Einstein condensation. To show that this distribution carries a Bose–Einstein condensation, subject to certain conditions on the phonon frequency distribution, we pass to the limit $N \to \infty$. There we assume that

(a) the phonon frequencies are distributed over a finite range $\omega_{\min} \leqslant \omega \leqslant \omega_{\max}$ according to a continuous distribution function, $\rho(\omega)$ defined by the condition that, for any sub-interval I of that range, $\int_I \rho(\omega)\,\mathrm{d}\omega$ is the limit, as $N \to \infty$, of N^{-1} times the number of modes with frequency in that range; and

(b) $\rho(\omega)$ tends to zero sufficiently fast, as $\omega \to \omega_{\min}$, to ensure that

$$\int_{\omega_{\min}}^{\omega_{\max}} \mathrm{d}\omega \rho(\omega)/(\omega - \omega_{\min}) \text{ is finite.} \tag{130}$$

Under these conditions, the standard theory† of condensation in an ideal Bose gas can be adapted to the present model as follows.

In the limit $N \to \infty$, the condition (126) reduces to

$$\mu \geqslant \omega_{\min}, \tag{131}$$

and the cases $\mu > \omega_{\min}$ and $\mu = \omega_{\min}$ have to be treated separately. In the former case, the occupation number, $n(\omega)$, for a mode of frequency ω may be obtained from (124) in the form

$$n(\omega) = \frac{1 + g^{-1}\mathrm{e}^{\beta\omega}}{\exp \beta(\omega - \mu) - 1}, \tag{132}$$

and, by condition (a), the self-consistency relation (125) reduces to the form

$$s = \phi(\mu), \tag{133}$$

where

$$\phi(\mu) = \int_{\omega_{\min}}^{\omega_{\max}} \mathrm{d}\omega\, \rho(\omega) \frac{(1 + g^{-1}\mathrm{e}^{\beta\omega})}{\exp \beta(\omega - \mu) - 1}. \tag{134}$$

It follows from this equation that $\phi(\mu)$ decreases monotonically from $\phi(\omega_{\min})$ to 0 as μ goes from $\omega_{\min}$ to ∞; and also, in view of (130), that $\phi(\omega_{\min})$ is finite. Hence, the condition (133) is fulfilled, for a unique value of μ, provided that

$$s < s_0 = \phi(\omega_{\min}). \tag{135}$$

If $s > s_0$, on the other hand, the condition (133) cannot be satisfied by a value of μ that exceeds $\omega_{\min}$, and therefore, in this case, we are forced, by (131), to the conclusion that $\mu = \omega_{\min}$. The formula (132) for $n(\omega)$ is then applicable only for $\omega > \omega_{\min}$ since the denominator there vanishes when

† See Landau and Wilde (1979) for a rigorous treatment of the ideal Bose gas, that can also be employed to prove condensation in the present model.

$\omega = \omega_{\min}$. Also, the contribution to the photon density, $\lim_{N\to\infty} N^{-1} \sum n_j$, due to the modes of frequency $> \omega_{\min}$, is then given by $\phi(\omega)$ $(= \phi(\omega_{\min}))$, as defined by (134). Thus, denoting by ν_0 the contribution to the phonon density due to the lowest frequency mode, we infer from (122) and the condition $\nu = s$ that

$$s = \nu_0 + \phi(\omega_{\min}),$$

i.e. by (135),

$$\nu_0 = s - s_0, \tag{136}$$

which is positive, as $s > s_0$. Hence we have a Bose–Einstein condensation with a finite density $(s - s_0)$ of phonons in the lowest frequency mode. This corresponds to a macroscopic excitation of this mode, with $\sim N(s - s_0)$ quanta, for a large but finite version of the model. Further, this condensation can occur *only* when $s > s_0$ since it requires that $\mu = \mu_{\min}$ which, by the above analysis, implies (136) with $\nu_0 \geqslant 0$.

We therefore conclude that the model exhibits a phase transition, corresponding to a Bose–Einstein condensation, when the pump parameter s passes beyond the threshold value s_0.

7.5. Concluding remarks

The models we have examined in this chapter provide illustrations of a very general phenomenon, namely the conversion of randomly supplied energy into ordered structures of macroscopic systems far from equilibrium.

However, the general picture of non-equilibrium ordering and phase transitions is much less developed than its equilibrium counterpart. One essential reason for this is that there is no known general characterization of stable non-equilibrium, states in open system, that can be employed to provide a unified description of them, in the way that the KMS fluctuation–dissipation or the global stability condition does for equilibrium states. Consequently, whereas the theory of order and phase transitions at equilibrium can be based on energy-versus-entropy arguments, there is no corresponding general methodology available once one departs from thermal equilibrium. Presumably, a prerequisite for such a methodology would be a classification scheme for the dynamical effects induced by reservoirs in open systems, that might serve to characterize their resultant stable states in much the same way as the global stability condition does for equilibrium states.

Appendix

Statistics of the forces $\chi(t)$

In this Appendix, we shall derive first the formulae (56)–(60) for the statistical properties of the stochastic forces $\chi(t)$ of the atomic model of

§7.2.1, and then corresponding properties (84)–(87) of the fluctuating forces acting on the atoms of the laser model of §7.3.

A.1. The atomic model

To obtain the statistical properties of the forces $\chi(t)$ of this model, we first observe that the following anticommutation relations between the fluctuating forces $\eta(t)$, $\eta^*(t)$ and the electronic Heisenberg operators $b(t')$, $b^*(t')$ follow from eqns (32) and (40), and the canonical anticommutation relations.

$$\{\eta_{j,\pm}(t), b_k(t')\} = 0 \tag{A.1}$$

$$\{\eta_{j,\pm}(t), b_k^*(t')\} = \begin{cases} \frac{1}{2}\gamma_{j,\pm}\delta_{jk} & \text{for} \quad t' = t \\ 0 & \text{for} \quad t' < t \end{cases} \tag{A.2}$$

$$\{\eta_{j,\pm}(t), \eta_k^*, (t')\} = \gamma_{j,\pm}\delta_{jk}\delta(t - t') \tag{A.3}$$

$$\{\eta_{j,+}(t), \eta_{k,-}^*(t')\} = 0, \tag{A.4}$$

and the forces $\eta_{j,\pm}(t)$ $(j = 1, 2)$, at arbitrary times, all intercommute.

It follows from equation (A.2) that the formulae (53)–(55) for the χ's may be re-expressed in the forms

$$\chi_v(t) = \sum_{j=1,2} (b_j^*(t)\eta_{j,-}(t) - b_j(t)\eta_{j,+}^*(t)) + \text{Hermitian conjugate}, \tag{A.5}$$

$$\chi_\rho(t) = \sum_{j=1,2} (-1)^j(b_j^*(t)\eta_{j,-}(t) - b_j(t)\eta_{j,+}^*(t)) + \text{Hermitian conjugate}, \tag{A.6}$$

and

$$\chi_\sigma(t) = (b_2^*(t), \eta_{1,-}(t) - b_1(t)\eta_{2,+}^*(t)) + (\eta_{2,-}^*(t)b_1(t) - \eta_{1,+}(t)b_2^*(t)). \tag{A.7}$$

Since, by (22), the forces $\eta_{j,-}(t)$ and $\eta_{j,+}^*(t)$ annihilate the reservoir state vector Ψ, it follows from (A.5–7) that the forces $\chi(t)$ all have zero expectation values for the initial state $\psi \otimes \Psi$ of $(S + R)$, i.e. that

$$\langle \chi_v(t) \rangle = \langle \chi_\rho(t) \rangle = \langle \chi_\sigma(t) \rangle = 0, \tag{A.8}$$

as in (56). Similarly since, by (30), (32), and the canonical anticommutation relations, the force $\eta(t)$ anticommutes with the zero-time operators b, b^* and therefore commutes with v, ρ, and σ, as defined by (44)–(46), it follows from (A.5–7) that

$$\langle v\chi_v(t) \rangle = \langle \rho\chi_\rho(t) \rangle = \langle \sigma^*\chi_\sigma(t) \rangle = 0, \tag{A.9}$$

as in (57).

Turning now the autocorrelation functions for the χ's we infer from

(A.5) and the fact that both $\eta_{j,-}(t)$ and $\eta^*_{j,+}(t)$ annihilate Ψ that

$$\langle \chi_\nu(t)\chi_\nu(t')\rangle = \sum_{j,k=1,2} \langle (b_j^*(t)\eta_{j,-}(t) - b_j(t)\eta^*_{j,+}(t))$$

$$\times (\eta^+_{k,-}(t')b_k(t') - \eta_{k,+}(t')b_k^*(t'))\rangle$$

$$\equiv \sum_{j,k=1,2} \langle b_j^*(t)\eta_{j,-}(t)\eta^*_{k,-}(t')b_k(t')\rangle \quad \text{(a)}$$

$$+ \sum_{j,k=1,2} \langle b_j^*(t)\eta_{j,-}(t)\eta_{k,+}(t')b_k^*(t')\rangle \quad \text{(b)}$$

$$- \sum_{j,k=1,2} \langle b_j(t)\eta^*_{j,+}(t)\eta_{k,-}(t')b_k(t')\rangle \quad \text{(c)}$$

$$- \sum_{j,k=1,2} \langle b_j^*(t)\eta_{j,-}(t)\eta_{k,+}(t')b_k^*(t')\rangle. \quad \text{(d)}$$

Now,

$$\text{term (a)} \equiv \sum_{j,k=1,2} [\langle b_j^*(t)\{\eta_{j,-}(t), \eta^*_{k,-}(t')\}b_k(t')\rangle$$

$$- \langle b_j^*(t)\eta^*_{k,-}(t')\{\eta_{j,-}(t), b_k(t')\}\rangle$$

$$+ \langle b_j^*(t)\eta^*_{k,-}(t')b_k(t')\eta_{j,-}(t)\rangle],$$

and therefore, by the anticommutation relations (A.1), (A.3), together with the fact that $\eta_{j,-}(t)$ annihilates the reservoir state vector Ψ,

$$\text{term (a)} = \sum_{j=1,2} \gamma_{j,-}\langle b_j^*(t)b_j(t)\rangle\delta(t-t').$$

Likewise, term (b) $= \sum_{j=1,2} \gamma_{j,+}\langle b_j(t)b_j^*(t)\rangle\ \delta(t-t')$. Hence, as $b_j(t)b_j^*(t) = 1 - b_j^*(t)b_j(t)$, it follows from these last two equations, together with the formulae (45) and (49), that

$$\text{term (a)} + \text{term (b)} = \gamma(1-\tau\bar{\rho}(t))\ \delta(t-t'),$$

which is equal to the RHS of (58). Therefore, in order to establish this latter formula, it suffices for us to show that the terms (c) and (d) in the above equation for the autocorrelation function of χ_ν both vanish. This we do as follows.

By the anticommutivity of $\eta^*_{j,+}(t)$ with $\eta^*_{k,-}(t)$,

$$\text{term (c)} = \sum_{j,k=1,2} \langle b_j(t)\eta^*_{k,-}(t')\eta^*_{j,+}(t)b_k(t')\rangle.$$

Further, by (A.2), $\eta^*_{j,+}(t)$ anticommutes with $b_k(t')$ for $t>t'$, and $\eta^*_{k,-}(t')$ anticommutes with $b_j(t)$ for $t<t'$. Therefore, as $\eta_{k,-}(t')$ and $\eta^*_{j,+}(t')$ both annihilate the reservoir state vector Ψ, it follows from the last equation that term (c) $=0$ if $t\neq t'$. On the other hand, if $t'=t$, then it follows from that last equation, together with the anticommutation relation (A.2) and the

annihilation of Ψ by $\eta^*_{j,+}(t)$ and $\eta_{j,-}(t')$ that term (c) $= \frac{1}{4}\sum_{j=1,2}\gamma_{j,+}\gamma_{j,-}$. Thus

$$\text{term (c)} = \begin{cases} \dfrac{1}{4}\sum_{j=1,2}\gamma_{j,+}\gamma_{j,-} & \text{if} \quad t=t' \\ 0 & \text{if} \quad t\neq t' \end{cases},$$

which implies that this term vanishes when integrated against any continuous function of t and t', i.e. that it is zero in the sense of distributions and so may be discarded. A more heuristic way of saying this to remark that while the terms (a), (b), and (c) all vanish for $t\neq t'$, the sum of the first two of these is infinite at $t=t'$, while (c) is finite there and so effectively makes no contribution when added to (a) and (b). Similarly, one can easily show that term (d) is zero in the sense of distributions and so can be discarded. Hence we conclude that the autocorrelation function $\langle\chi_v(t)\chi_v(t')\rangle$ is given by the above expression for the sum of the terms (a) and (b) and so satisfies eqn (58).

The formulae (59) and (60) for the autocorrelation functions for χ_ρ and χ_σ may be obtained by the same procedure.

A.2. The laser model

As a first step towards obtaining the means and autocorrelation functions of the forces $\chi(t)$ of the laser model, Σ_N, we formulate some of its key algebraic properties, analogous to those given by (A.1)–(A.4) for the single atom. Thus, we start by expressing the Hamiltonian for Σ_N in the form

$$H = H^{(0)} + H^{FA} \tag{A.10}$$

where $H^{(0)}$ is the sum of the Hamiltonians for the subsystems $(F+R(F))$, $(A_1+R(A_1)),\ldots,(A_N+R(A_N))$, consisting of the field mode F and the atoms $A_1,\ldots,A_N$, in interaction with their respective reservoirs, and H^{FA} represents the coupling between atoms and field mode, and is given by (70). We then express the Heisenberg operators for the time-translations of the observables, Q, of Σ_N in the form

$$Q(t) \equiv \exp(\mathrm{i}Ht)Q\exp(-\mathrm{i}Ht) = V(t)\tilde{Q}(t)V^{-1}(t), \tag{A.11}$$

where

$$\tilde{Q}(t) = \exp(\mathrm{i}H^{(0)}t)Q\exp(-\mathrm{i}H^{(0)}t) \tag{A.12}$$

is the evolute of Q in the absence of any coupling between the atoms and the field, and

$$V(t) = \exp(\mathrm{i}Ht)\exp(-\mathrm{i}H^{(0)}t), \tag{A.13}$$

which is given formally by the Dyson expansion

$$V(t) = I + \sum_1^\infty \mathrm{i}^n \int_0^t \mathrm{d}t_1 \ldots \int_0^{t_{n-1}} \mathrm{d}t_n \tilde{H}^{FA}(t_1)\ldots\tilde{H}^{FA}(t_n), \tag{A.14}$$

where, by (70) and (A.12),

$$\tilde{H}^{FA}(t) = \lambda \sum_{1}^{N} (\tilde{a}^*(t)\tilde{b}^*_{l,1}(t)\tilde{b}_{l,2}(t) + \tilde{a}(t)\tilde{b}^*_{l,2}(t)\tilde{b}_{l,1}(t)). \tag{A.15}$$

In view of the general definition (A.12) of $\tilde{Q}(t)$, we see that $\tilde{a}(t)$ and $\tilde{b}_{l,j}(t)$ are the Heisenberg operators for the mode F and the atom A_l, each in interaction with its own reservoir only. Hence, it follows from our specifications in §7.3 of the algebraic structure of the model that

(a) for each atom A_l, the fermion operators $\tilde{b}_{l,j}(t)$ and $\eta_{l,j,\pm}(t)$ of $(A_l + R(A_l))$ satisfy the anticommutation relations analogous to (A.1–4); and

(b) these operators anticommute with all the fermion operators for any other atom and its reservoir, $(A_m + R(A_m))$, and commute with $\tilde{a}(t)$ and $\tilde{a}^*(t)$.

It is now a simple consequence of (a), (b), and eqn (A.15) that the commutator $[\tilde{H}^{FA}(t'), \eta_{j,l,\pm}(t)]$ vanishes for $t' < t$ and reduces to a well-defined operator ($\tilde{a}(t)$ or $\tilde{a}^*(t)$ times a bounded operator) if $t' = t$. This implies, by (A.14), that $\eta_{l,j,\pm}(t)$ commutes with $V(t')$ for $t' \leqslant t$ and consequently, by (a), (b), and (A.11), that the stochastic forces $\eta_{l,j,\pm}(t)$ and the Heisenberg operators $b_{l,j}(t)$ possess the following algebraic properties.

(a)′ For each atom A_l, the operators $b_{l,j}(t)$ and $\eta_{l,j,\pm}(t)$ satisfy the anticommutation relations (A.1–4); and

(b)′ for different atoms, A_l and A_m, $\eta_{l,j,\pm}(t)$ anticommutes with the stochastic forces $\eta_{m,k,\pm}(t')$ and $\eta^*_{m,k,\pm}(t')$ and also, for $t \geqslant t'$, with the Heisenberg operators $b_{m,k}(t')$ and $b^*_{m,k}(t')$.

In view of these properties, it is now a straightforward matter to derive the formulae (84)–(87) for the means and correlations of the forces $\chi_{\nu,l}$, $\chi_{\rho,l}$ and $\chi_{\sigma,l}$ by the method used in §A.1 to obtain the corresponding formula for the single atom.

References

AIZENMANN, M. (1980). Translational invariance and instability of phase coexistence in the two-dimensional Ising model, *Commun. Math. Phys.* **74,** 83–94.

ALCANTARA, J. and DUBIN, D. A. (1981). I*-algebras and their applications, *Publ. R.I.M.S., Kyoto University* **17,** 179–99.

ARAKI, H. (1969). Gibbs states of a one-dimensional quantum system, *Commun. Math. Phys.* **14,** 120–7.

——— and SEWELL, G. L. (1977). Local thermodynamical stability and the KMS conditions of quantum lattice systems, *Commun. Math. Phys.* **52,** 103–9.

——— and WOODS, E. J. (1963). Representations of the canonical commutation relations describing a nonrelativistic free Bose gas, *J. Math. Phys.* **4,** 637–62.

BAKER, G. (1972). Ising model with a scaling interaction, *Phys. Rev. B* **5,** 2622–33.

BATEMAN, H. and ERDELYI, A. (1955). *Higher transcendental functions*. McGraw-Hill, New York.

BAXTER, R. J. (1972). Partition functions of the eight-vertex lattice model, *Annals of Physics* **70,** 193–228.

BLEHER, P. M. and SINAI, YA. G. (1973). Investigation of the critical point in models of the type of Dyson's hierarchical model, *Commun. Math. Phys.* **33,** 23–42.

——— and ——— (1975). Critical indices for Dyson's asymptotically hierarchical models, *Commun. Math. Phys.* **45,** 247–78.

BOGOLIUBOV, N. N. (1962). *Phys. Abh. S.U.* **1,** 229.

BOHM, D. and PINES, D. (1951). A collective description of electron interactions I. Magnetic Interactions, *Phys. Rev.* **82,** 625–34.

BORCHERS, H. J. (1962). On the structure of the algebra of field operators, *Nuovo Cimento* **24,** 214.

——— (1973). Algebraic aspects of Wightman field theory in statistical mechanics and quantum field theory, in *Statistical mechanics and field theory* (Ed. R. N. Sen and C. Weil). Israel University Press, Jerusalem/London.

BRATTELI, O. and ROBINSON, D. W. (1979). 'Operator algebras and quantum statistical mechanics Vol. 1; (1981), Vol. 2. Springer, Heidelberg/New York.

BRYDGES, D. and FEDERBUSH, P. (1980). Debye screening, *Commun. Math. Phys.* **73,** 197–246.

CALLEN, H. (1960). *Thermodynamics*. John Wiley, London/New York/Sydney.

——— and WELTON, T. A. (1951). Irreversibility and generalised noise, *Phys. Rev.* **83,** 34–40.

CAPOCACCIA, D., CASSANDRO, M., and OLIVIERI, E. (1974). A study of metastability in the Ising model, *Commun. Math. Phys.* **39,** 185–204.

CASSANDRO, M. and JONA-LASINIO, G. (1978). Critical point behaviour and probability theory, *Advances in Physics* **6,** 913–41.

CHANDRASEKAR, S. (1961). *Hydrodynamic and hydromagnetic stability*. Oxford University Press, Oxford.

CHOQUET, G. (1966). *Topology*. Academic Press, London/New York.

COOK, J. M. (1953). The mathematics of second quantisation, *Trans. Am. Math. Soc.* **74,** 224–45.

DAVIES, E. B. (1974). Markovian master equations, *Commun. Math. Phys.* **39,** 91–110.

DELL'ANTONIO, G. F., DOPLICHER, D., and RUELLE, D. (1966). A theorem on canonical commutation and anticommutation relations, *Commun. Math. Phys.* **2,** 223–30.

DI CASTRO, C., and JONA-LASINIO, G. (1969). On the microscopic foundations of scaling laws, *Phys. Lett.* **29A,** 322–3.

DICKE, R. H. (1954). Coherence in spontaneous radiation processes, *Phys. Rev.* **93,** 99–110.

DIEUDONNÉ, J. (1960). Foundations of modern analysis. Academic Press, London/New York.

DIRAC, P. A. M. (1958). *Principles of quantum mechanics*, 4th edn. Clarendon Press, Oxford.

DOBRUSHIN, R. L. (1972). Gibbsian state which describes co-existence of phases for a three-dimensional Ising model, *Theory Prob. Appl.* **17,** 582.

DOMB, C., and GREEN, M. S. (1972). *Phase transitions and critical phenomena*, Vol. 1. Academic Press, London/New York.

DUBIN, D. A. (1974). *Solvable models in algebraic statistical mechanics*. Clarendon Press, Oxford.

——— and SEWELL, G. L. (1970). Time-translations in the algebraic formulation of statistical mechanics, *J. Math. Phys.* **11,** 2990–8.

DYSON, F. J. (1969). Existence of a phase transition in a one-dimensional ferromagnet', *Commun. Math. Phys.* **12,** 91–107.

——— and LENARD, A. (1967). Stability of matter, *J. Math. Phys.* **8,** 423–434.

——— and ——— (1968). *J. Math. Phys.* **9,** 698–711.

——— and LIEB, E. H., and SIMON, B. (1978). Phase transitions in quantum spin systems with isotropic and non-isotropic interactions, *J. Stat. Phys.* **18,** 335–83.

EMCH, G. G. (1966). The definition of states in quantum statistical mechanics, *J. Math. Phys.* **7,** 1413–20.

——— (1967*a*). Common interpretations of phase transitions in various models, *J. Math. Phys.* **8,** 13–18.

——— (1967*b*). Van der Waals wiggles, Maxwell rule and temperature-dependent excitations, *J. Math. Phys.* **8,** 19–25.

——— (1972). *Algebraic methods in statistical mechanics and quantum field theory.* John Wiley, New York/London.

——— and KNOPS, H. J. F. (1970). Pure thermodynamical phases as extremal KMS States', *J. Math. Phys.* **11,** 3008–318.

——— and RADIN, C. (1971). Relaxation of local thermal deviations from equilibrium, *J. Math. Phys.* **12,** 2043–6.

FANNES, M., and VERBEURE, A. (1977). Correlation inequalities and equilibrium states, *Commun. Math. Phys.* **55,** 125–31; **57,** 165–71.

FISHER, M. E. (1964). The free energy of a macroscopic system, *Arch. Rat. Mech. Anal.* **17,** 377–410.

——— (1967*a*). The theory of condensation and the critical point, *Physics* **3,** 255–83.

——— (1967*b*). The theory of equilibrium critical phenomena, *Repts Prog. Phys.* **30,** 615–730.

—— (1974). The renormalisation group in the theory of critical behaviour, *Rev. Mod. Phys.* **46,** 597–616.

Fock, V. (1932). Konfigurationraum und Zweite Quantelung, *Z. Phys.* **75,** 622–47.

Fröhlich, H. (1968). Long-range correlations and energy storage in biological systems, *Int. J. Quantum Chem.* **2,** 641–9.

—— (1969). Quantum mechanical concepts in biology, pp. 13–22 of *Theoretical physics and biology* (Ed. M. Maurois). North Holland, Amsterdam.

—— (1973). The connection between macro- and microphysics, *Rivista del Nuovo Cimento* **3,** 490–534.

—— (1980). The biological effects of microwaves and related questions, *Advances in Electronics and Electron Physics* **53,** 85–152.

—— (1983). Evidence for coherent excitations in biological systems, *Int. J. Quantum Chem.* **23,** 1589–95.

Fröhlich, J., Simon, B., and Spencer, T. (1976). Infra-red bounds, phase transitions and continuous symmetry breaking, *Commun. Math. Phys.* **50,** 79–85.

—— and Lieb, E. H. (1978). Phase transitions in anisotropic lattice systems, *Commun. Math. Phys.* **60,** 233–67.

——, Israel, R. B., Lieb, E. H., and Simon, B. (1978). Phase transitions and reflection positivity I. General theory and long range lattice models, *Commun. Math. Phys.* **62,** 1–34.

——, ——, ——, —— (1980). Phase transitions and reflection positivity II. Lattice systems with short range and Coulomb interactions, *J. Stat. Phys.* **22,** 297–347.

Gallavotti, G. (1972). The phase separation line in the two-dimensional Ising model, *Commun. Math. Phys.* **27,** 103–36.

——, Miracle-Sole, S., and Robinson, D. W. (1968). Analyticity properties of the anisotropic Heisenberg model, *Commun. Math. Phys.* **10,** 311–24.

Gelfand, I., and Naimark, M. A. (1943). On the imbedding of normed rings into the ring of operators in Hilbert space, *Mat. Sborn., N.S.* **12,** 197–217.

Ginibre, J. (1965). Reduced density matrices of quantum gases, *J. Math. Phys.* **6,** 238–51; 252–62; 1432–66.

—— (1968). Reduced density matrices of the anisotropic Heisenberg model, *Commun. Math. Phys.* **10,** 140–54.

—— (1969). Existence of phase transitions for quantum lattice systems, *Commun. Math. Phys.* **14,** 205–34.

Ginzburg, V. L. (1955). On the theory of superconductivity, *Nuovo Cimento* **2,** 1234–50.

—— and Landau, L. D. (1950). On the theory of superconductivity, *Zh. Eksberlin i Teor. Fyz.* **20,** 1064–82 (in Russian).

Glansdorff, P. and Prigogine, I. (1971). *Thermodynamic theory of structure, stability and fluctuations*, Wiley–Interscience, London/New York.

Glauber, R. J. (1963*a*). Time-dependent statistics of the Ising model, *J. Math. Phys.* **4,** 294–307.

—— (1963*b*). The quantum theory of optical coherence, *Phys. Rev.* **130,** 2529–39.

Graham, R., and Haken, H. (1970). Laser light – first example of a second-order phase transition far away from equilibrium, *Z. Phys.* **237,** 31–46.

Griffiths, R. B. (1964). Peierls proof of spontaneous magnetisation in a two-dimensional Ising ferromagnet, *Phys. Rev.* **136A,** 437–9.

——— (1966). Spontaneous magnetization in idealised ferromagnets, *Phys. Rev.* **152,** 240–246.
——— (1967). Correlations in Ising ferromagnets I, *J. Math. Phys.* **8,** 478–83.
——— (1981). What's wrong with real-space renormalisation group transformations, pp. 463–79 of *Random fields,* Vol. 1 (Ed. J. Fritz, J. L. Lebowitz, and D. Szasz). North Holland, Amsterdam/Oxford/New York.
HAAG, R. (1961). Canonical commutation relations in quantum field theory and functional integration, pp. 353–81 of *Lectures in Theoretical Physics*, Vol. 3 (Ed. W. E. Brittin, B. W. Downs, and J. Downs). Interscience, New York/London.
——— and KASTLER, D. (1964). An algebraic approach to quantum field theory, *J. Math. Phys.* **5,** 846–61.
HAAG, R., HUGENHOLTZ, N. M., and WINNINK, M. (1967). On the equilibrium states in quantum statistical mechanics, *Commun. Math. Phys.* **5,** 215–36.
HAKEN, H. (1970). *Handbuch der Physik*, Bd. XXV/2C. Springer, Heidelberg/Berlin/New York.
——— (1975). Cooperative phenomena in systems far from thermal equilibrium and in non-physical systems, *Rev. Mod. Phys.* **47,** 67–121.
HEPP, K. (1972). Quantum theory of measurement and macroscopic observables, *Helv. Phys. Acta* **45,** 237–48.
——— and LIEB, E. H. (1973). Phase transitions in reservoir-driven open systems, with applications to lasers and super-conductors, *Helv. Phys. Acta* **46,** 573–603.
HERTEL, P., and THIRRING, W. (1971). Thermodynamical instability of a system of gravitating fermions, pp. 310–23 of *Quanten und Felder* (Ed. H. P. Durr) Vieweg, Braunschweig.
———, NARNHOFER, H., and THIRRING, W. (1972). Thermodynamic functions for fermions with gravostatic and electrostatic interactions, *Commun. Math. Phys.* **28,** 159–76.
HOPF, E. (1942). Abzweigung einer periodischen Lösung von einer Stationären Lösung eines Differentialsystems, *Ber. Math-Phys. Kl. Sächs. Akad. Wiss. Leipzig* **94,** 1–22.
HUGENHOLTZ, N. M., and WIERINGA, J. D. (1969). On locally normal states in quantum statistical mechanics, *Commun. Math. Phys.* **11,** 183–97.
JAUCH, J. M. (1968). *Foundations of quantum mechanics*, Addison Wesley, Reading, Mass./London.
JONA-LASINIO, G. (1975). The renormalisation group: a probabilistic view, *Nuovo Cimento* B **26,** 99–119.
JOSEPHSON, B. D. (1964). Coupled superconductors, *Rev. Mod. Phys.* **36,** 216–20.
KADANOFF, L. P. (1966). Scaling laws for Ising models near T_c, *Physics* **2,** 263–72.
KADISON, R. V. (1965). Transformations of states in operator theory and dynamics, *Topology* **3,** Suppl. 2, 177–98.
KHINCHIN, A. I. (1949). *Mathematical foundations of statistical mechanics.* Dover, New York.
KOSSAKOWSKI, A., FRIGERIO, A., GORINI, V., and VERRI, M. (1977). Quantum detailed balance and the KMS condition, *Commun. Math. Phys.* **57,** 97–110.
KOTECKY, R., and SHLOSMAN, S. B. (1982). First-order phase transitions in large entropy lattice models, *Commun. Math. Phys.* **83,** 493–515.
KUBO, R. (1957). Statistical mechanical theory of irreversible processes, I, *J. Phys. Soc. Japan* **12,** 570–86.
LANDAU, L. D., and LIFSHITZ, E. M. (1959). *Statistical physics*, Pergamon, London/New York/Paris.

Landau, L. J. and Wilde, I. (1979). On the Bose–Einstein condensation of the ideal gas, *Commun. Math. Phys.* **70,** 43–51.

Lanford, O. E., and Robinson, D. W. (1968*a*). Mean entropy of states in quantum statistical mechanics, *J. Math. Phys.* **9,** 1120–5.

——— and ——— (1968*b*). Statistical mechanics of quantum spin systems III, *Commun. Math. Phys.* **9,** 327–38.

——— and Ruelle, D. (1969). Observables and infinity and states with short range correlations in statistical mechanics, *Commun. Math. Phys.* **13,** 194–215.

Lebowitz, J. L. (1968). Statistical mechanics – a review of selected rigorous results, *Annual Review of Physical Chemistry* **19,** 389–418.

——— and Penrose, O. (1966). Rigorous treatment of the Van der Waals theory of the liquid–vapor transition, *J. Math. Phys.* **7,** 98–113.

——— and ——— (1968). Analytic and clustering properties of thermodynamics functions and distribution functions for classical lattice and continuum systems, *Commun. Math. Phys.* **11,** 99–124.

——— and Martin-Löf, A. (1972). On the uniqueness of the equilibrium state for Ising spin systems, *Commun. Math. Phys.* **25,** 276–82.

Lieb, E. H. (1966). Quantum-mechanical extension of the Lebowitz–Penrose theorem on the Van der Waals theory, *J. Math. Phys.* **7,** 1016–24.

——— (1967*a*). Exact solution of the F model of an antiferroelectric, *Phys. Rev. Lett.* **18,** 1046–8.

——— (1967*b*). Exact solution of the two-dimensional Slater KDP model of a ferroelectric, *Phys. Rev. Lett.* **19,** 108–10.

——— (1976). The stability of matter, *Rev. Mod. Phys.* **48,** 553–69.

——— and Ruskai, M. B. (1973). Proof of the strong subadditivity of quantum-mechanical entropy, *J. Math. Phys.* **14,** 1938–41.

——— and Thirring, W. (1975). Bound for the kinetic energy of fermions which proves the stability of matter, *Phys. Rev. Lett.* **35,** 687–9.

Lukacs, E. (1970). *Characteristic functions*. Griffin, London.

Martin, Ph. A. (1977). Stochastic dynamics of Ising models, *J. Stat. Phys.* **16,** 149–68.

Martin, P. C., and Schwinger, J. (1959). Theory of many-particle systems: I, *Phys. Rev.* **115,** 1342–73.

Mayer, J. E. (1947). Integral equations between distribution functions of molecules, *J. Chem. Phys.* **15,** 187–201.

Mermin, N. D., and Wagner, H. (1966). Absence of ferromagnetism or antiferromagnetism in one- or two-dimensional isotropic Heisenberg models', *Phys. Rev. Lett.* **17,** 1133–6.

Messager, A., and Miracle-Sole, S. (1975). Equilibrium states of a two-dimensional Ising model in the two-phase region, *Commun. Math. Phys.* **40,** 187–96.

Messer, J. (1981). On the gravitational phase transition in the Thomas–Fermi model, *J. Math. Phys.* **22,** 2910–17.

Miracle-Sole, S., and Robinson, D. W. (1970). Statistical mechanics of quantum particles with hard cores II: The equilibrium states, *Commun. Math. Phys.* **19,** 204–18.

Narnhofer, H. (1970). On Fermi lattice systems with quadratic Hamiltonians, *Acta Phys. Austriaca* **31,** 349–53.

——— and Sewell, G. L. (1980). Equilibrium states of gravitational systems, *Commun. Math. Phys.* **71,** 1–28.

NIEMEYER, TH., and VAN LEEUWEN, J. M. J. (1974). Wilson theory for two-dimensional Ising spin systems, *Physica* **71,** 17–40.

NYQUIST, H. (1928). Thermal agitation of electric charge in conductors, *Phys. Rev.* **32,** 110–13.

ONSAGER, L. (1944). Crystal statistics, I. A two-dimensional model with an order–disorder transition, *Phys. Rev.* **65,** 117–49.

OSTERWALDER, K. and SCHRADER, R. (1973). Axioms for euclidean Green's functions *Commun. Math. Phys.* **31,** 83–112.

——— and ——— (1975). Axioms for euclidean Green's functions II *Commun. Math. Phys.* **42,** 281–305.

PEIERLS, R. E. (1936). On Ising's model of ferromagnetism, *Proc. Camb. Phil. Soc.* **32,** 477–81.

PENROSE, O., and ONSAGER, L. (1956). Bose–Einstein condensation and liquid helium, *Phys. Rev.* **104,** 576–84.

——— and LEBOWITZ, J. L. (1971). Rigorous treatment of metastable states in the Van der Waals–Maxwell theory, *J. Stat. Phys.* **3,** 211–36.

PIRIGOV, S. A., and SINAI, YA. G. (1975). Phase diagrams of classical lattice systems, *Theor. Math. Phys.* **25,** 1185–92.

——— and ——— (1976). Phase diagrams of classical lattice systems, continuation, *Theor. Math. Phys.* **26,** 39–49.

POWERS, R. T. (1971). Self-adjoint algebras of unbounded operators, *Commun. Math. Phys.* **21,** 85–124.

——— (1974). *Trans. Am. Math. Soc.* **187,** 261–93.

RADIN, C. (1977). The dynamical instability of nonrelativistic many-body systems, *Commun. Math. Phys.* **54,** 69–79.

RIESZ, F., and SZ-NAGY, B. (1955). *Functional analysis*. F. Ungar, New York.

ROBERTS, A. W., and VARBERG, D. E. (1973). *Convex functions*. Academic Press, London/New York.

ROBINSON, D. W. (1968). Statistical mechanics of quantum spin systems II, *Commun. Math. Phys.* **7,** 337–48.

——— (1969). A proof of the existence of phase transitions in the anisotropic Heisenberg model, *Commun. Math. Phys.* **14,** 195–204.

——— (1971). The thermodynamical pressure in quantum statistical mechanics, *Lecture Notes in Physics*, Vol. 9. Springer, Berlin/Heidelberg/New York.

RUELLE, D. (1967). A variational formulation of equilibrium statistical mechanics and the Gibbs phase rule, *Commun. Math. Phys.* **5,** 324–9.

——— (1969*a*). *Statistical mechanics*. W. A. Benjamin Inc., New York.

——— (1969*b*). Symmetry breakdown in statistical mechanics, pp. 169–94 of *Cargese lectures*, Vol. 4 (Ed. D. Kastler). Gordon & Breach, New York.

SCHIFF, L. (1968). *Quantum mechanics*, 3rd edn. McGraw-Hill, New York.

SCHRÖDINGER, E. (1952). *Statistical thermodynamics*. Cambridge University Press.

SCHWEBER, S. (1961). *An introduction to relativistic quantum field theory*. Row & Peterson, Evanston, Illinois.

SEGAL, I. E. (1947). Postulates for general quantum mechanics, *Annals of Mathematics* **48,** 930–48.

——— (1951). A class of operator algebras which are determined by groups, *Duke Math. J.* **18,** 221–65.

SEWELL, G. L. (1970). Unbounded local observables in quantum statistical mechanics, *J. Math. Phys.* **11,** 1868–84.

——— (1974). Relaxation, amplification and the KMS conditions, *Annals of Physics* **85,** 336–77.

——— (1976). Metastable states of classical systems, *Annals of Physics* **97,** 55–79.
——— (1977*a*). KMS conditions and local thermodynamical stability of quantum lattice systems II, *Commun. Math. Phys.* **55,** 53–61.
——— (1977*b*). Metastable states of quantum lattice systems, *Commun. Math. Phys.* **55,** 63–6.
—— (1980). Stability, equilibrium and metastability in statistical mechanics, *Phys. Rep.* **57,** 307–42.
——— (1982). W*-dynamics of infinite quantum systems, *Lett. Math. Phys.* **6,** 209–13.
SIMON, B., and SOKAL, A. D. (1981). Rigorous energy–entropy arguments, *J. Stat. Phys.* **25,** 679–94.
SINAI, YA. G. (1976). Automodal probability distributions, *Teor. Veroyatn. Primen* **21,** 63–78((in Russian).
——— (1982). *Theory of phase transitions; rigorous results*. Pergamon, Oxford/New York/Toronto/Sydney/Paris/Frankfurt.
STREATER, R. F. (1967). The Heisenberg ferromagnet as a quantum field theory, *Commun. Math. Phys.* **6,** 233–247.
THIRRING, W. (1980). *Quantum mechanics of large systems*. Springer, New York/Vienna.
THOULESS, D. (1969). Long-range order in one-dimensional Ising systems, *Phys. Rev.* **187,** 732–3.
TISZA, L. (1966). *Generalised thermodynamics*. M.I.T. Press, Cambridge, Mass., and London.
UHLMANN, A. (1962). Über die Definition der Quantenfelder nach Wightman und Haag, *Z. K.-Marx U. Leipzig* **11,** 213–7.
VAN BEIJEREN, H. (1975). Interface sharpness in the Ising system, *Commun. Math. Phys.* **40,** 1–7.
VANHEUVERZWIJN, P. (1979). Metastable states in the infinite Ising model, *J. Math. Phys.* **20,** 2665–70.
VAN HOVE, L. (1955). Quantum mechanical perturbations giving rise to a statistical transport equation, *Physica* **21,** 517–40.
VON NEUMANN, J. (1931). Die Eindeudigkeit der Schrödingerschen Operatoren, *Math. Annalen* **104,** 570–8.
——— (1955). *Mathematical foundations of quantum mechanics*. Princeton University Press.
WEISS, K., and HAUG, H. (1973). Bose condensation and superfluid hydrodynamics', pp. 219–35 of *Cooperative phenomena*, (Ed. H. Haken and M. Wagner). Springer, Berlin/Heidelberg/New York.
WHITTAKER, E. T., and WATSON, G. N. (1946). A course of modern analysis. Cambridge University Press.
WIDOM, B. (1965). Equation of state in the neighbourhood of the critical point, *J. Chem. Phys.* **43,** 3898–905.
WIGNER, E. P. (1959). *Group theory and its application to quantum mechanics of atomic spectra*. Academic Press, New York.
WILSON, K. G. (1971). Renormalisation group and critical phenomena, *Phys. Rev.* B **4,** 3174–83; 3184–205.
——— and FISHER, M. E. (1972). Critical exponents in 3.99 dimensions', *Phys. Rev. Lett.* **28,** 240–3.
——— and KOGUT, J. (1974). The renormalisation group and the ε-extension, *Phys. Rep.* **12,** 75–200.
WINNINK, M. (1973). Some general properties of thermodynamic states in the

algebraic approach, 311–33 of *Statistical mechanics and field theory* (Ed. R. N. Sen and C. Weil). Israel University Press, Jerusalem/London.

YANG, C. N. (1962). Concept of off-diagonal long range order and the quantum phases of liquid He and of superconductors, *Rev. Mod. Phys.* **34,** 694–704.

YOSIDA, K. (1965). *Functional analysis*. Springer, Heidelberg/Berlin/New York.

Index

Note: The letter 'n' following a page number denotes that the entry appears in a footnote.